Working with
Animal Chromosomes

Working with Animal Chromosomes

Herbert C. Macgregor

Department of Zoology,
University of Leicester, Leicester, UK

and

Jennifer M. Varley

Department of Zoology,
University of Leicester, Leicester, UK
and
Department of Medical Biochemistry,
University of Geneva, Geneva, Switzerland

Chapter 7 contributed by

Aimee Bakken

Department of Zoology,
University of Washington,
Seattle, Washington, USA.

R. Stephen Hill

Department of Genetics,
University of Liverpool,
Liverpool, UK

A Wiley–Interscience Publication

JOHN WILEY & SONS

Chichester · New York · Brisbane · Toronto · Singapore

Library of Congress Cataloging in Publication Data:
  Macgregor, Herbert C.
  Working with animal chromosomes.
  'A Wiley–Interscience publication.'
  Includes bibliographical references and index.
  1. Chromosomes. 2. Animal genetics—Technique.
I. Varley, Jennifer M. II. Title.
QH600.M33 1983   591.87′322 82–23788

ISBN 0 471 10295 4

British Library Cataloguing in Publication Data:

Macgregor, Herbert C.
  Working with animal chromosomes.
  1. Chromosomes 2. Cytology—Techniques
I. Title II. Varley, Jennifer M.
  574.87′332   QH600

ISBN 0 471 10295 4

Printed at The Pitman Press, Bath, Avon.

To 'Mick'
(H. G. Callan)

who taught us how to work with chromosomes

Contents

Preface

In 1942 the late C. D. Darlington FRS and L. F. La Cour published the first edition of their well known book *The Handling of Chromosomes*. In that book they described chromosomes as being 'amongst the largest molecules whose chemistry is being pursued . . . the smallest living structures whose movements and transformations can be seen and followed', and of course they attached great importance to the study of chromosome form and function. Their book has for years been a valuable source of information on how to work with chromosomes from the standpoint of a microscopist and it will doubtless continue to serve its useful purpose for many years to come. Since 1942 there have been remarkable advances in biological science and biotechnology stretching right through to the most recent exciting developments in the fields of microtechnique, light and electron microscopy and recombinant DNA. Bioscience is everywhere reaching right down to the level of molecules and molecular interactions. For the most part, supramolecular structures have been described and our understanding of many of them is good. Now the biologist tends to be more concerned with systems and events representing the interaction of molecules within defined structural environments.

Chromosomes, because they are the embodiment of the genome, are very much at the centre of all this and interest in them has expanded beyond what might have been even the wildest dreams of cytologists in the Darlingtonian era. Ways of handling them have diversified and become complicated and technically challenging. Nonetheless chromosome cytology and cytogenetics are still very much sciences of seeing, businesses of manipulation and microscopy, and in that sense the content of this book is not fundamentally different from that of Darlington and La Cour's 40 years ago.

Our primary objective has been to describe in detail the ways of working with chromosomes that are most useful for tackling current problems relating to the organization, function and behaviour of the genome in eukaryotic cells. Our descriptions and protocols have in so far as is possible been written with three different kinds of person in mind. In the first place we have tried to offer something to the teacher both at senior high school and university undergraduate level who wants to construct a useful, exciting and inexpensive laboratory practical exercise. So we have provided basic protocols that work for the preparation of all kinds of chromosomes, employing materials that are readily available and quite easily handled. Secondly we have tried to provide enough information in each section to allow an able but inexperienced investigator to learn a particular technique and develop skills that will lead into an original research programme. In effect we hope that this book will help and inspire more people to learn how to handle lampbrush chromosomes, prepare chromatin, visualize transcription, locate genes on polytene chromosomes, identify fine scale polymorphisms on mitotic chromosomes, and so on. Lastly we have tried to provide something useful for the experienced investigator who simply wants to use a particular specialized chromosome technique to answer one specific question quickly and effectively, or who wants to give something a try to see if it would be useful and appropriate to move a research programme in that direction. With regard to each of the systems that we have covered we have tried to produce a model protocol for a method that we know to work reproducibly with the materials that we specify. Naturally these model techniques will have to be adjusted or modified here and there to suit other materials.

Plant biologists will doubtless wonder why we have so studiously avoided reference to any methods for dealing with plant chromosomes. The reasons are simple. In the first place there are several excellent and up-to-date texts on methods in plant cytology. The book entitled *Chromosome Techniques, Theory and Practice* (Sharma and Sharma, 1972) is particularly noteworthy. Secondly, many of the methods that we describe are directly applicable to plant material provided that steps are taken to deal with the problem of the plant cell wall, and where this is not so it is most often because plants do not have chromosomes of the kind that are under consideration: no plants have polytene chromosomes and only one unicellular alga is known to exhibit anything remotely resembling a lampbrush chromosome. Finally, and perhaps most importantly, neither of us has ever worked with plant chromosomes, and we have considered it important in this book to offer guidance and advice only on matters of which we have good first hand experience.

We wish to thank the following colleagues for data, methods, recipes, references, encouragement, and advice:

Dr Mike Ashburner, Genetics Department, University of Cambridge, Cambridge, UK.

Mr George Breckon, Medical Research Council Radiobiology Unit, Harwell, UK.

Professor H. G. Callan, FRS, Department of Zoology, University of St Andrews, St Andrews, UK.

Dr S. C. R. Elgin, Department of Biology, Washington University, St Louis, Missouri, USA.

Dr J. G. Gall, Department of Biology, Yale University, New Haven, Connecticut, USA.

Dr John Gosden, Medical Research Council Clinical Population Cytogenetics Unit, Western General Hospital, Edinburgh, UK.

Dr Gareth Jones, Department of Genetics, University of Birmingham, Birmingham, UK.

Dr James Kezer, Department of Biology, University of Oregon, Eugene, Oregon, USA.

Dr Garry Morgan, Zoology Department, University of Washington, Seattle, Washington, USA.

Dr C. Redfern, Department of Molecular Biology, University of Edinburgh, Edinburgh, UK.

Dr John Savage, Medical Research Council Radiobiology Unit, Harwell, UK.

Dr M. Seabright, Wessex Regional Cytogenetics Unit, General Hospital, Salisbury, UK.

Dr G. B. White, Entomology Department, London School of Hygiene and Tropical Medicine, London, UK.

We also thank Christopher Macgregor, Pat Pellatt, and Jacqueline Boswell for their help with the production of the drawings, photographs, and typescript. A number of persons have generously provided us with photographs, and these will be acknowledged individually in the text. Finally, we thank the University of Leicester for providing us over the years with excellent facilities with which to apply and develop the techniques that we describe, and we offer special thanks to Joyce Moody, Lesley Barnett, and Frances Barker. Last but not least our gratitude goes to Jennifer and Graham, for all their help and support while this work was in progress.

December 1982   H.C.M. and J.M.V.

References

Darlington, C. D. and La Cour, L. F. (1942). *The Handling of Chromosomes*, George Allen and Unwin Ltd., London.

Sharma, A. K. and Sharma, A. (1972). *Chromosome Techniques, Theory and Practice*, Butterworth, London.

1

Chromosome size, number, and shape

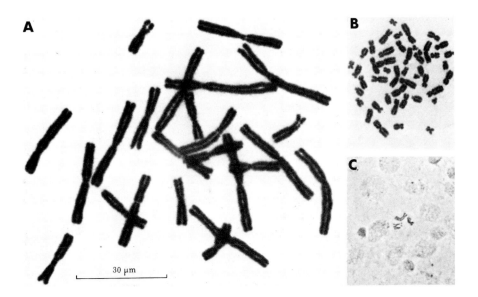

Chromosome sets from the crested newt (*Triturus cristatus*) (A), man (B), and *Drosophila melanogaster* (C), each photographed and reproduced at the same magnification. The two tiny fourth chromosomes of *Drosophila melanogaster* can be seen as a pair of dots just to the left of centre in the space between the three pairs of larger chromosomes. [Reproduced with the kind permission of the Royal Society from Macgregor, H. C., *Phil.Trans.Roy.Soc.Lond.*, **B283**, 309–318 (1978)]

The study of chromosomes nearly always involves manipulation and microscopy and measurement. In this sense it is an immensely satisfying occupation since it encourages manual dexterity, presents visual images, and generates hard data that has real molecular significance. Naturally there are well established conventions for describing and measuring chromosomes, and it seems appropriate to begin by reviewing these and at the same time exploring some of the principles behind the objective approach to chromosome work.

Table 1.1   Haploid chromosome numbers

Species	Haploid chromosome number
Homo sapiens	23
Drosophila melanogaster	4
D. virilis	6
Chironomus tentans	4
C. plumosus	6
Locusta migratoria	12/13(X0/XX)
Xenopus laevis	18
X. tropicalis	10
Rana pipiens	13
Triturus cristatus	12
Notophthalmus viridescens	11
Plethodon cinereus	14
Batrachoseps attenuatus	13
Mus musculus	20
Cricetulus griseus (Chinese hamster)	11
Mesocricetus auratus (Golden hamster)	22
Gallus gallus domesticus (includes microchromosomes)	78

The haploid chromosome number for a species is generally referred to by the letter n; diploid being $2n$, tetraploid $4n$, and so on. The range of chromosome numbers in animals is wide. At one end of the spectrum is a fully authenticated haploid number of 2 in the iceryine coccid from Mexico, *Steatococcus tuberculatus*. At the other extreme is a lycaenid butterfly with a haploid chromosome number estimated to be between 217 and 223. Table 1.1 shows the haploid numbers for a range of animals that commonly feature in chromosome studies. Other chromosome numbers can be found in Michael White's book *Animal Cytology and Evolution* (1973), in the book *Cytotaxonomy and Vertebrate Evolution* edited by Chiarelli and Capanna (1973), or in the chromosome atlases compiled by Hsu and Benirschke (e.g. Hsu and Benirschke, 1968; Benirschke and Hsu, 1971).

It is of the utmost importance in chromosome studies to take note of chromosome size since this has far reaching significance in relation to evolution, speciation, and chromosome organization. Indeed far too few students and teachers attach importance to actually measuring chromosomes and representing them as objects of definitive sizes that are characteristic of the

Table 1.2   *C* values

Species	DNA per haploid chromosome set (pg)
Homo sapiens	3.5
Drosophila melanogaster	0.14
Xenopus laevis	3.0
X. tropicalis	1.5
Triturus cristatus	23
Notophthalmus viridescens	35
Plethodon cinereus	20
Necturus maculosus	95
Mus musculus	3.5
Gallus gallus domesticus	1.6

species from which they came. Since all of the organism's genomic DNA resides in its chromosomes, chromosome size and number will reflect the size of the whole genome. Genome size is normally expressed as grams of DNA per haploid chromosome set, often referred to as the '*C* value' of the organism. The amount of DNA per diploid cell is then the '2*C* value'. These terms should be applied carefully and critically. The term *genome* refers to all the DNA embodied in the haploid set of chromosomes. *C* refers to the amount of DNA expressed in picograms (i.e. grams $\times 10^{-12}$) per haploid set of chromosomes. Some *C* values are given in Table 1.2, from which it will be seen, for example, that the human genome incorporates nearly 20 times as much DNA as that of *Drosophila melanogaster*, and that of an average urodele has nearly 10 times that of a man. Such wide differences are most striking when chromosomes are drawn or photographed at comparable magnifications or when the amount of DNA they contain is expressed in terms of the total length of linear DNA duplex (introductory illustration). The significance of these differences is most clear when one begins to think about the genome in terms of DNA sequences that may or may not be transcribed and translated into messenger RNA and protein, or when we consider the physical problems of condensing and moving chromosomes of vastly different sizes within the sort of time scale that is normal for animal mitoses. The differences in size between, for example, the fourth chromosome of *Drosophila melanogaster* and the largest chromosomes in the karyotypes of certain plethodontid salamanders is comparable in scale to the difference between a salmon and a nuclear submarine!

The measurement of *C* values usually involves a combination of biochemical and optical techniques. Microdensitometric measurement of the amount of Feulgen dye bound to nuclei or chromosomes is the commonest approach, usually complemented by some kind of biochemical estimate of the amount of DNA extracted from a known number of cell nuclei. More recently flow cytophotometric measurement of nuclear DNA contents has become popular on account of the accuracy of the method and large number of nuclei that can be measured in a very short time. More information on these techniques is given in Chapter 10.

The measurement and description of chromosomes in mitotic or meiotic metaphase is simple but should be carried out with critical regard to potential sources of error and with strict reference to established conventions. Generally speaking, a chromosome in mitotic metaphase has only two distinguishing features: its length and a transverse constriction that marks the position of the centromere. In certain circumstances it is possible to introduce other distinguishing marks by differential staining with Giemsa or quinacrine and procedures for doing this are detailed in Chapter 4. From the chromosome length and centromere position three factors can be calculated: the centromere index, the arm ratio, and the relative length of the chromosome. The first two factors, centromere index and arm ratio, tell us about the chromosome itself. The relative length tells us about the size of the chromosome in relation to other chromosomes in the set.

> *Centromere index* is defined as *the length of the shorter of the two chromosome arms multiplied by 100 and divided by the length of the whole chromosome*. It is therefore expressed as a percentage.
> *Arm ratio* is defined as *the length of the longer arm of the chromosome divided by the length of the shorter one*. It is therefore always greater than 1.
> *Relative length* is defined as *the length of the whole chromosome multiplied by 100 and divided by the total length of all the chromosomes in the haploid set, including the one being measured*. Once again, it is expressed as a percentage.

It is worth noting that the relative length can be used as a rough guide to the proportion of the genome that is represented by a particular chromosome, but in this respect one must remember that it is a linear measurement and not a measure of volume or mass. The internationally agreed terminology that is applied to the description of chromosomes is shown in Table 1.3 and discussion of the use of these terms can be found in Volpe and Gebhardt (1968) and Levan *et al.* (1964). Full details of conventions applying to chromosomes that have been banded with Giemsa or fluorochromes, with particular regard to human chromosomes, are given in the Proceedings of the 1971 Paris Conference on Standardization in Human Cytogenetics (Paris Conference, 1971, 1975).

There are several ways of measuring chromosomes. All of them involve having well spread microscope preparations in which the chromosomes appear with good contrast and sharp outlines. The simplest, fastest, and most

Table 1.3 Chromosome classification by centromere position

Centromere position	Alternative terminology	Chromosome symbol	Centromere index range†
Nearly median	Metacentric	m	46–49
Submedian	Submetacentric (more metacentric)	sm	36–45
Submedian	Submetacentric (less metacentric)	sm	26–35
Subterminal	Acrocentric	st	15–30

$$\dagger\text{Centrometre index} = \frac{\text{Length of short arm}}{\text{Chromosome relative length}} \times 100.$$

straightforward method is to draw the chromosomes at the highest possible magnification with the help of a *camera lucida* or drawing apparatus attached to the microscope. When the outlines of the chromosomes and their centromere positions have been carefully recorded the slide is replaced by a micrometer slide etched in divisions of 10 μm and, with the microscope and *camera lucida* set up exactly as they were for drawing the chromosomes, a scale line of appropriate length is drawn alongside the drawing of the chromosome set. The length of each chromosome or chromosome arm can then be measured using a piece of string, a simple map measuring wheel, or, fastest and most sophisticated of all, a digitizer or graphics tablet linked to a microcomputer. Chromosome arm lengths can then easily be converted into absolute units by reference to the scale bar on the drawing. Other methods involve photographing the chromosomes first at the highest magnification that will produce clear images that can subsequently be enlarged by projection and printing for measurement.

The first step in the characterization of a complete chromosome set is usually referred to as karyotyping. It is normally followed by measurement of each chromosome and preparation of a diagrammatic representation of the entire set in an idiogram. Both these procedures begin with a search for good well spread mitotic metaphases or early second meiotic anaphases. In the latter there are usually two nicely arranged and virtually identical haploid sets of chromosomes, and a useful double checking of chromosome length and identity is therefore possible (Figure 1.1). It is then necessary to determine most carefully the chromosome number of the animal under study. Is the chromosome set constant from cell to cell and from one individual to another? Are there supernumerary chromosomes that vary in number within or between individuals or populations? Are there any heteromorphic sex chromosomes? Which sex is heterogametic? Whilst it is desirable to settle such matters before embarking on a karyotyping operation, it is nonetheless entirely possible that the process of attempting to produce a karyotype and the detailed searching and recording that is involved may expose idiosyncra-

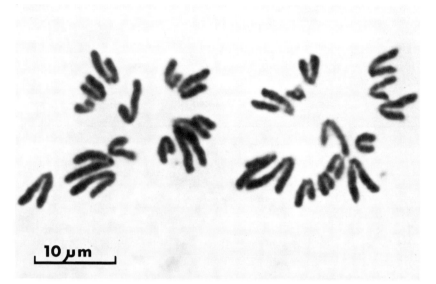

Figure 1.1   A photomicrograph of a squash preparation of a well spread
second meiotic anaphase from the red-backed salamander, *Plethodon
cinereus*. This kind of preparation is sometimes especially useful for
karyotype analysis because the dimensions of each chromosome can be
checked against those of its counterpart in the other set. The disadvantage
of the second meiotic anaphase, of course, is that it does not show
heterozygosities or heteromorphisms that may be important features of
the animal's chromosome set. [Reproduced with permission from Kezer,
J. & Macgregor, H. C., *Chromosoma (Berl.)*, **33**, 146–166 (1971)]

sies in the chromosome set that were not apparent in the course of a preliminary examination. It should be noted that many features of a chromosome
set can *only* be determined by careful examination of mitotic metaphases
with occasional back-up evidence from examination of first meiotic prometaphase or metaphase figures.

Having established confidence in the chromosome number and identified
any unusual features of the chromosome set, the chromosomes can then be
drawn, photographed, and measured, and centromere indices, arm ratios,
and relative lengths can be determined. Finally a karyotype and an idiogram
are produced after the fashion shown in Figures 1.2–1.4.

The production of a karyotype requires a good high magnification photographic print of a complete chromosome set with the minimum number of
overlapping chromosomes. If there are overlaps in the photograph then make
some extra prints so that it will be possible to make cut-out images of each
individual chromosome. The original photomicrographs of the chromosomes
should be taken with a high resolution oil-immersion objective. Then enlarge
the images during photographic printing so that the smallest chromosome in

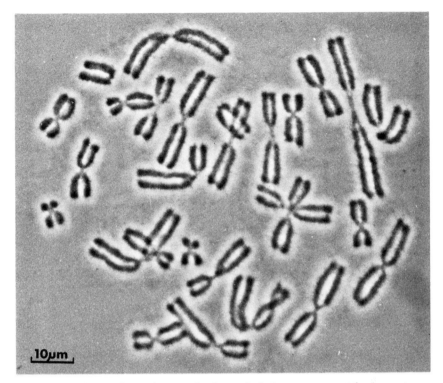

Figure 1.2 A photomicrograph of a typical chromosome set from a gut epithelial cell of the North American salamander, *Aneides ferreus*. There are several places in this set where the chromosomes overlap or touch. Two or three prints would therefore be needed in order to provide cut-outs of all the chromosomes. [Reproduced with permission from Mac-gregor, H. C. and Jones, C., *Chromosoma (Berl.)*, **63**, 1–9 (1977)]

the set is easily measureable with as much accuracy as the sharpness of its outline permits. With a pair of scissors, cut out each chromosome from the enlarged print, trimming around the chromosome without removing any part of its image. Arrange the cut-out chromosomes in order of decreasing length from left to right on a piece of mounting board. Then, with regard to their lengths and centromere positions and any other regular features, attempt to arrange the chromosomes in homologous pairs. This will be quite simple if the chromosomes have been differentially stained with Giemsa or fluoro-chromes or if they have highly diverse shapes. It will be much more difficult if the chromosomes form a more or less continuous series in terms of length and are similar with regard to their centromere positions. In such a case, any attempt to arrange the chromosomes in homologous pairs has to be partly based on guesswork. Once the chromosomes are arranged in pairs and in

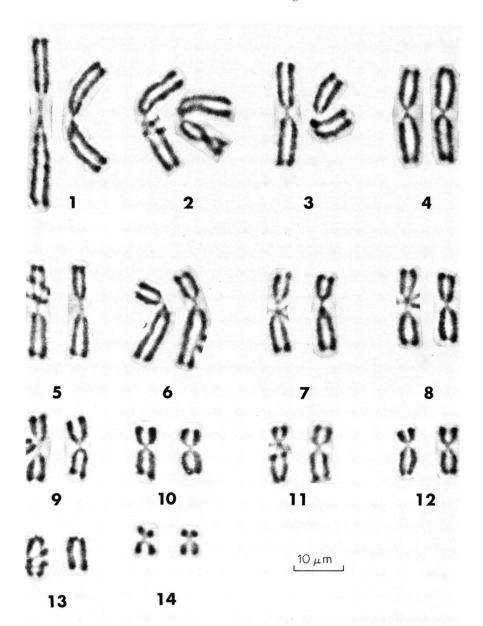

Figure 1.3   The chromosomes of the set shown in Figure 1.2 set out in pairs and in order of decreasing length in so far as is possible without recourse to special techniques of chromosome identification such as Giemsa banding. Note that all the chromosomes except number 13 are meta- or submetacentric. Number 13 is acrocentric.

order of length, stick them down to the mounting board in a series of rows, always with the short arm uppermost and the centromeres lined up along each row (see Figures 1.2 and 1.3). Number the chromosomes and indicate the final magnification with a scale bar. The karyotype is now complete.

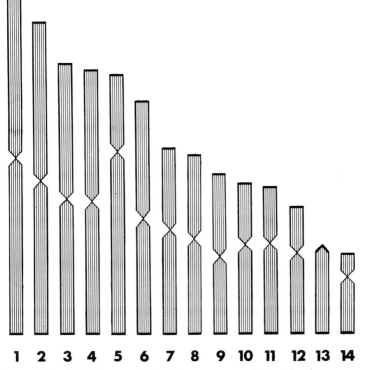

1 2 3 4 5 6 7 8 9 10 11 12 13 14

Figure 1.4 An idiogram of the chromosomes of *Aneides ferreus* constructed from data obtained from the preparation shown in Figure 1.3, recorded, processed, and printed out in this form with an Apple computer and graphics tablet, utilizing a program obtainable from Leicester Computer Centre (see Appendix 1)

For the production of an idiogram it is necessary to draw or photograph and measure the chromosomes, and then determine centromere indices, arm ratios, and relative lengths. Finally, construct an idiogram in the manner shown in figure 1.4, in which each chromosome is represented as a thick vertical bar having a length in accordance with the chromosome's relative length, and the position of the centromere and any other conspicuous features are clearly and precisely indicated. The vertical bars are then arranged in descending order of length with the longest on the left and the shortest on the right and in every case with the shorter arm of the chromosome uppermost. Idiograms should not be constructed on the basis of data derived from a single preparation but should represent mean values for chromosome and arm lengths from measurements of a substantial number of chromosome sets taken from several individuals of the species concerned. Most importantly, the idiogram should not be attempted until the observer has become fully familiar with the chromosome set of the organism. Only in this way is it possible to avoid overlooking important oddities like precocious or asynchronous condensation of certain chromosome arms, variations in the lengths of centromeric or secondary constrictions, and variations in chromosome number.

The general principles of karyotyping can be applied effectively to chromosomes from any cell type provided that they can be adequately visualized. Lampbrush chromosomes (see Chapter 6) provide excellent material for accurate karyotyping so long as it is possible to identify centromeres. Chromosomes in first meiotic prophase can also be accurately karyotyped by employing the technique developed by Counce and Meyer (1973) and so successfully exploited by Moses and his associates over the past 10 years (see Moses *et al.*, 1975, 1977, 1982) in which synaptonemal complexes are spread on an air/water interface, picked up on a microscope slide, and then visualized by silver staining.

As the use and availability of microcomputers becomes more and more common, programs are rapidly becoming available for processing data from chromosome sets. One such program is described by Green *et al.* (1980). Another simple program specifically designed for the production of idiograms from photomicrographs of chromosome sets using an Apple computer and an Apple graphics tablet, the commonest and most inexpensive digitizing equipment currently available, can be obtained from The Leicester Computer Centre Ltd (see Appendix 1).

References

Benirschke, K., and Hsu, T. C. (1971). *Chromosome Atlas: Fish, Amphibians, Reptiles and Birds*, Springer Verlag, Berlin, Heidelberg, and New York.

Chiarelli, A. B., and Capanna, E. (eds) (1973). *Cytotaxonomy and Vertebrate Evolution*, Academic Press, London and New York.

Counce, S. J., and Meyer, G. F. (1973). Differentiation of the synaptonemal complex and the kinetochore in *Locusta* spermatocytes studied by whole mount electron microscopy. *Chromosoma (Berl.)*, **44**, 231–253.

Green, D. M., Bogart, J. P., and Anthony, E. H. (1980). An interactive, microcomputer-based karyotype analysis system for phylogenetic taxonomy. *Comput. Biol. Med.*, **10**, 219–227.

Hsu, T. C., and Benirschke, E. (1968). *Atlas of Mammalian Chromosomes*, Springer Verlag, Berlin, Heidelberg, and New York.

Levan A., Fredga, K., and Sandberg, A. A. (1964). Nomenclature for centromeric position on chromosomes. *Hereditas (Lund)*, **52**, 201–220.

Moses, M. J., Counce, S. J., and Paulson, D. F. (1975). Synaptonemal complex complement of man in spreads of spermatocytes, with details of the sex chromosome pair. *Science*, **187**, 363–365.

Moses, M. J., Slatton, G. H., Gambling, T. M., and Starmer, C. F. (1977). Synaptonemal complex karyotyping in spermatocytes of the Chinese hamster (*Cricetulus griseus*). *Chromosoma (Berl.)*, **60**, 345–375.

Moses M. J., Poorman, P. A., Roderick, T. H., and Davisson, M. T. (1982). Synaptonemal complex analysis of mouse chromosomal rearrangements. IV. Synapsis and synaptic adjustment in 2 paracentric inversions. *Chromosoma (Berl.)*, **84**, 457–474.

Paris Conference (1971). *Standardization in Human Cytogenetics* (D. Bergsma, ed), National Science Foundation, Washington, D.C.

Paris Conference (1975). *Standardization in Human Cytogenetics*, Supplement (D. Bergsma, ed), National Science Foundation, Washington, D.C.

Volpe, P. E., and Gebhardt, B. M. (1968). Somatic chromosomes of the marine toad, *Bufo marinus* (Linne). *Copeia* **1968** (3), 570–576.

White, M. J. D. (1973). *Animal Cytology and Evolution*, 3rd edn, Cambridge University Press, London and New York.

2

Mitotic chromosomes

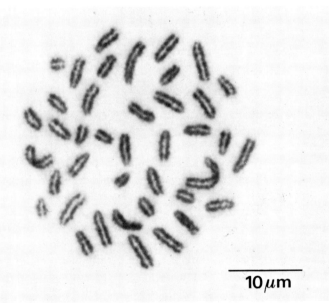

A mouse metaphase chromosome complement. Magnification 5000×.
[This photomicrograph was kindly provided by C. V. Beechy, G. Brec-
kon, and A. H. Cawood, Medical Research Council Radiobiology Unit,
Harwell]

In this chapter methods will be described for the preparation of mitotic
chromosomes using a technique that we will refer to as the splash technique
or simply as 'air drying'. Details of the stages of mitosis will not be described
since most standard texts on genetics or cytology cover these adequately. For

the majority of the cell cycle time, the chromosomes are in an interphase condition where the DNA is diffuse and no details of the karyotype can be determined. However, during metaphase the chromosomes are maximally condensed and have quite distinctive forms according to the species from which they are derived. Karyotype descriptions are nearly always produced from examination of mitotic metaphase chromosomes, as are banding patterns and localizations of DNA sequences by *in situ* nucleic acid hybridization. There are two principal ways of preparing mitotic chromosomes for microscopy. The first is the production of a cell suspension in which a reasonable proportion of the cells are dividing, followed by the preparation of metaphase spreads by splashing the fixed cell suspension on to a microscope slide. The second major technique is the squash technique and this will be discussed extensively in Chapter 3. The concern of this chapter will be with the preparation of well spread mitotic metaphase figures from a wide variety of species and tissue types using the splash technique.

In the first section of this chapter the basic principles of the techniques will be described. These principles will be explained in the order in which they are used during the preparation of metaphase chromosomes and not in the chronological order of their development, although it may be interesting to the reader to note the latter. For an excellent history of mammalian cytogenetics, the reader is referred to an entertaining account by Hsu (1979). Without a doubt the majority of work has been carried out on mammalian systems, mainly due to the medical importance of cytogenetic studies in man and in animals of domestic importance to man.

The second section comprises a detailed description of a method for the preparation of human chromosomes from leucocytes which can be taken as a model for the 'splash' or 'air drying' technique. The protocol is one which gives good reproducible results with a high yield of well spread metaphase chromosome preparations. The following section will describe a number of variations on the technique and should allow the model procedure to be modified accordingly and applied to any mammalian species. A method is also presented for preparing metaphase chromosomes from a variety of different tissues. It is hoped that the reader will thus be presented with a range of techniques which are within the facilities and capabilities of the complete novice, the teacher attempting a class practical, or a potential researcher into mammalian cytogenetics.

More specialized protocols follow in Section IV, which comprises techniques for preparing good metaphase spreads from a wide variety of non-mammalian species. It is hoped that the description of these techniques will cover a sufficiently wide spectrum to allow them to be adapted to suit material from any vertebrate species.

The final section in this chapter covers four techniques for preparing mitotic chromosomes by squashing the relevant tissue rather than by air drying. Details of the squash technique are given in Chapter 3, but it is relevant to discuss here those applications of the squash technique that are specifically concerned with the preparation of mitotic material.

I.   Principles of the splash technique

The obvious prerequisite for the examination of mitotic chromosomes is a source of dividing material. There are three types of material which can be used. The first is material that is actively dividing *in vivo*, so that *in vitro* cultivation of the tissue is unnecessary. An example of such material is bone marrow, a useful source of dividing tissue in mammals, amphibians, and birds. Other examples include the spleen, intestinal epithelium and fibroblasts. This type of material has two main advantages. In the first place it is fast. Culturing is unnecessary and chromosome preparations can be made immediately. Secondly, there is no problem of artefacts of the kind that are often produced during the culturing of cells. Unfortunately only a limited number of tissue types are dividing rapidly enough to be of use and even in these the number of actively dividing cells may be small. The second type of material suitable for splashing is that which divides relatively infrequently in the body, but which can be set up in long term culture. The advantages of using this type of material are that many tissue types can be used, and the cells that grow up will survive through many population doublings without karyotype changes and so are available for further study. Moreover, tissue samples may be stored at 4°C in culture medium for several weeks before setting up the long term cultures. Many tissue types can be set up in long term culture, including those which actively divide *in vivo* and others such as heart, kidney, and liver. It is not intended to give details of tissue culture in this book and the reader is instead referred to one of many sources of cell culture techniques (e.g. Kruse and Patterson, 1973; Jakoby and Pastan, 1979). In general, techniques for initiating cell cultures involve removing a piece of the relevant tissue using sterile technique, chopping it aseptically and then briefly treating it with a protease such as trypsin, collagenase, or hyaluronidase. (A useful review of this kind of methodology is given by Bashor, 1979.) Protease treatment dissociates the tissue into single cells or small clumps of cells. After washing the cell suspension to remove the protease, the cells are cultured in a suitable medium and allowed to grow until a useful number of cells are present (see Ham and McKeehan, 1979, for a review). Usually a culture period of 1–3 weeks will be adequate to produce many dividing cells, the time depending on the cell type. The cells are then arrested in metaphase by the addition of a spindle inhibitor and are detached from the culture vessel using trypsin. This method has superceded a much less efficient method of culturing small blocks of tissue in a plasma clot and examining the cells in the small zone of outgrowth from these blocks.

The splash technique can also be used with a third type of material, that which is nucleated but which does not normally divide *in vivo*. The obvious and common example of this type of tissue is that of leucocytes in the peripheral blood film of mammals, birds, and reptiles. Under normal circumstances leucocytes do not undergo mitotic division, but they can be stimulated to do so by the addition of a mitogen (for a review see Cunningham *et al.*, 1975). The importance of the splash technique cannot be overemphasized. It arose because of the necessity for a simple, quick, and reliable method for obtaining human chromosome preparations and the technique has now been adapted for use with many other systems. Short term *in vitro* cultures of peripheral blood are initiated in medium supplemented with the mitogen and are cultured for 48–72 h. There are a number of commercially available mitogens which will induce normally non-dividing cells to undergo DNA synthesis and mitotic division. The most commonly used mitogen is phytohaemagglutinin (PHA), an extract of the red kidney bean, *Phaseolus vulgaris*. Phytohaemagglutinin was initially isolated because of its property of agglutinating erythrocytes and therefore assisting in the isolation of leucocytes, but it was discovered also to possess a mitogenic activity (Nowell, 1960; Barker, 1969). Phytohaemagglutinin M is the form that is generally used; it is purchased as a lyophilized powder, reconstituted, and used according to the manufacturer's instructions. A useful alternative to PHA is pokeweed mitogen, isolated from *Phytolacca americana* root (Farnes *et al.*, 1964). Pokeweed mitogen has very similar mitogenic properties to PHA, but has lower leucagglutinating properties and it can be used for samples which respond poorly to PHA. There is, however, evidence that exposure of the skin to pokeweed mitogen can lead to circulation of mitotic cells in the peripheral bloodstream, although there have been no reports of other long term effects. It should, however, be used with care and skin exposure avoided. Both of the above mitogens can be used for a wide variety of animal mononuclear cells *in vitro*, but the most commonly used is PHA.

Once a good source of dividing tissue has been identified it is advisable to arrest the cells at metaphase using a spindle inhibitor. Three spindle inhibitors are in frequent use: colchicine, colcemid, and vinblastine sulphate. Colchicine and colcemid are closely related alkaloids with similar properties, although the mode of action of colchicine is the best characterized. The isolation of spindle inhibitors was first reported by plant cytologists and colchicine has been used to induce polyploidy in plant tissues as well as to arrest cell division at metaphase (Blakeslee and Avery, 1937; Levan, 1938; Eigsti and Dustin, 1955). Both colchicine and colcemid are isolated from *Colchicum* species and both have identical properties of spindle inhibition, but colcemid is less toxic to animal cells. The toxicity is not important if one is preparing mitotic tissue from cultured cell suspensions or monolayers, because the incubation time is short, but if the spindle inhibitor is injected into the whole animal to arrest mitoses, toxicity may then become an important factor. Separation of the chromosomes to the spindle poles is inhibited by colchicine or colcemid by their disruptive action on the microtubules. These alkaloids therefore serve a useful secondary purpose. Not only are the cells arrested at metaphase, but, because the chromosomes no longer lie on the equatorial plate, they separate well after fixation. However, treatment with a spindle inhibitor can cause excessive contraction of the metaphase chromosomes on prolonged incubation, producing so-called 'C-mitotic' figures. Therefore in some cases the treatment may have to be short, or even entirely omitted, in order to obtain less contracted chromosomes which will be of more use for fine structure analysis of the chromosomes. Although omission of the spindle inhibition step is possible, if a large number of metaphases are required its use is strongly recommended. The optimal time for this step usually has to be determined empirically, but an indication of the range of concentration and time will be given later in this chapter.

After culturing and/or inhibition of spindle formation, the cells are ready for harvesting. If they are in a monolayer, it is necessary to detach them from the culture vessel by a brief trypsin digestion. First, the culture medium is removed and replaced with either fresh medium containing trypsin or a balanced salt solution containing trypsin. A brief incubation is adequate to remove most cells and a rapid tap on the side of the culture vessel will remove any cells which are still adhering. This cell suspension must be washed to remove the trypsin before proceeding further with the protocol. This is done by spinning the cells in the medium plus trypsin in a centrifuge at about 100g, removing the supernatant, and replacing it with fresh medium minus trypsin. Cells growing in suspension (e.g. leucocytes or bone marrow cells) must also be centrifuged to remove the culture medium, but there is no need to respin these suspensions. It is important to remember that, with any cell culture, until the medium is removed colchicine will still be present, and this must be taken into account when timing the colchicine step.

The cells now undergo a hypotonic treatment to swell the cell and nucleus and allow better separation of chromosomes on splashing. Prior to 1952 most studies of mammalian cytogenetics involved embedding and sectioning testis material, a very lengthy process. In 1952 two independent groups made an important technical advance in the preparation of mammalian chromosomes (Hsu, 1952; Hughes, 1952). Instead of washing the cultures in the usual isotonic saline, Hsu fortuitously used a hypotonic saline solution and noted the swelling effect this had on the cell and nucleus. The second researcher (Hughes, 1952) independently discovered the same phenomenon during studies on the effects of hypotonicity on dividing cells. A hypotonic treatment is essential for mammalian and possibly also bird cells, but in some amphibian and invertebrate systems it can be omitted. A hypotonic step is routinely incorporated in the preparation of mitotic (and meiotic) chromosomes from any species, as invariably the results obtained are much better than if this step were omitted. In addition to the hypotonic treatment causing the cell and nucleus to swell, there are reports that interchromosomal connections are disrupted (Klášterská and Ramel, 1979), thus further improving the chances of good chromosome separation. It is important to achieve optimal conditions of hypotonic swelling: if swelling is excessive, the chromosomes will be so scattered that it may be impossible to identify whole metaphase sets; if it is insufficient, then the chromosomes will not separate and the metaphase complements will have many overlapping chromosomes and will be correspondingly difficult to analyse. The size of the chromosomes themselves can be a factor in determining the optimal hypotonic treatment. Cells from an animal with large chromosomes may need a longer hypotonic treatment than those from one with small chromosomes, but care must be taken because excessive hypotonic treatment may also cause some dispersion of the chromosomal material, leaving the chromosomes with a fuzzy appearance and making the resolution of banding patterns virtually impossible.

The cells are centrifuged after hypotonic treatment; the pellet is resuspended in a small amount of hypotonic solution and is then fixed. The fixative that is widely used is a mixture of methanol and acetic acid made up in the ratio of three parts methanol to one part of acetic acid (3 : 1). This fixative is made up immediately before use and can be used ice-cold or at room temperature. An alternative fixative is ethanol–acetic acid. Methanol is often preferred where ethanol is suspected of containing appreciable amounts of water or impurities. It is important that the fixative is fresh, as old 3 : 1 contains large amounts of methyl acetate and is inadequate as a fixative. Treatment with methanol–acetic acid has been shown to extract large amounts of protein, especially histones (Dick and Johns, 1968; Sumner *et al.*, 1973), but the DNA is largely native. It is possible to include formaldehyde as a fixative and this prevents some loss of histones and non-histone proteins, but the results are usually less satisfactory than those using methanol–acetic acid alone. One of the most important functions of the fixation step is the removal of much of the cell debris and proteinaceous material from the cell suspension, to produce a much cleaner preparation of chromosomes. Cells can be kept at 4°C in 3 : 1 methanol–acetic acid for a period of up to a few weeks, but it is recommended that the preparations are made as soon as possible after thorough fixation. Cells which have been stored in fixative must be washed in fresh fixative before proceeding to the final step, that of making the splash preparation.

The principle of the splash technique is that a suspension of fixed cells is dropped on to a microscope slide in such a way that the nuclei burst and the chromosomes are released. The chromosomes then become firmly attached to the slide as the fixative dries. There are many slight modifications to the technique of splashing the cell suspension, but for the most part these depend upon the degree of severity required to burst open the nuclei and produce a discrete group of well spread non-overlapping chromosomes. The techniques range from simply dropping some of the fixed cell suspension from a pipette on to a microscope slide to flaming the slides and causing the fixative to ignite. The range of possible techniques will be discussed in Sections II and III.

Immediately after splashing, the preparations should be examined either directly, using phase contrast optics, or after staining with an appropriate stain.

II. Preparation of human chromosomes from leucocytes using the splash technique

A. Equipment and materials

(a) Autoclave – not needed if plastic disposable material is used

(b) Small bench centrifuge for speeds up to 100*g*

(c) Centrifuge tubes – approximate capacity 15 ml

(d) 37°C incubator or water bath

(e) Microscope with phase contrast (preferable) and bright field (essential) condenser, plus phase contrast 16× and 40× objectives and bright field 63× and 100× objectives

(f) 1 ml and 20 ml sterile syringes and needles

(g) Autoclaved McCartney vials or sterile disposable 30 ml vials

(h) One heparinized vial, as above but containing 100 IU heparin

(i) Sterile 1, 5, and 10 ml pipettes

(j) Culture medium (e.g. RPMI 1640, HEPES-buffered, plus Glutamine Gibco-Biocult cat. no. 041-2400 or Hams F10, Gibco cat. no. 041-2390)

(k) Serum (heat-inactivated neonatal or newborn calf serum, virus and mycoplasma screened, Gibco cat. no. 023–6010)

(l) Phytohaemagglutinin M (Gibco cat. no. 061-0576)

(m) Colchicine, stock 1 mg/ml solution, stored at −20°C.

(n) 0.075 M KCl (hypotonic solution)

(o) 3 : 1 methanol–glacial acetic acid (made freshly immediately before use and mixed well)

(p) Antibiotics (a penicillin–streptomycin mixture, Gibco cat. no. 043-5140), optional

(q) Microscope slides

B. Procedure

(1) Collect 10–20 ml of peripheral blood by venepuncture (this must be carried out by a medically qualified person). Dispense into a sterile vial containing heparin at a concentration of approximately 10–20 IU/ml of blood and shake well to mix. Heparin is an anticoagulant and so prevents clotting of the blood; it may be obtained from most pharmacists.

(2) Allow the vial of whole blood to stand undisturbed at 37°C for 1–2 h. After this time, the blood will have separated into two layers, the bottom layer comprising mainly red blood cells and the upper layer containing an enrichment of leucocytes. After the blood has separated, dispense 0.8 ml aliquots of the upper layer aseptically into sterile 30 ml vials containing the following: 8 ml of medium (e.g. RPMI 1640); 2 ml of heat inactivated neonatal calf serum; 0.1 ml of phytohaemagglutinin M; 100 IU/ml penicillin, 100 µg/ml streptomycin, optional. Alternatively, 0.8 ml of whole blood may be inoculated into the above mixture directly after collection without separation into red and white cell layers.

(3) Incubate at 37°C for 72 h, shaking gently twice every 24 h to cut down on agglutination of erythrocytes. This is especially important if whole blood was used to inoculate the cultures, but red blood cells will also be present even if blood enriched for leucocytes is used. Agglutination of the red blood cells is mainly due to the action of the phytohaemagglutinin.

(4) For the final 1.5 h of culture time, add colchicine to a final concentration of 0.4 µg/ml of culture medium. There is no need to filter-sterilize the colchicine and aseptic technique need no longer be employed. Remember to include the time taken over the next centrifugation step in the timing of the colchicine treatment.

(5) Shake the culture to ensure that the cells are in suspension and decant the contents of each vial separately into a 15 ml centrifuge tube. Spin in a bench centrifuge at 100*g* for 10 min and remove and discard the supernatant using a Pasteur pipette.

(6) Resuspend the cell pellet by adding 10 ml of 0.075 M KCl which has been prewarmed to 37°C and drawing the cells gently in and out of a Pasteur pipette. Incubate at 37°C for 14 min, then spin at 100*g* for 6 min. Note that this gives a total time in the hypotonic solution of 20 min. This time can be adjusted to obtain the optimal conditions (see subsequent discussion).

(7) Remove the supernatant, leaving about 0.5 ml of fluid above the cell pellet. Resuspend the cell pellet in this fluid by sucking the cells gently in and out of a Pasteur pipette until a fine cell suspension with no large clumps of cells remains.

(8) Draw the resuspended cells into the Pasteur pipette and expel the suspension slowly into a fresh 15 ml centrifuge tube containing 10 ml of freshly prepared, ice-cold 3 : 1 fixative. Mix gently but thoroughly using the

Pasteur pipette. Addition of the cells to the fixative in this way prevents aggregation of the cells, which can be a problem if the fixative is added to the cell pellet. Leave the cell suspension for at least 10 min, during which time the erythrocytes will become completely disrupted.

(9) Spin in a bench centrifuge for 10 min at 100g, remove the supernatant, and add 10 ml of fresh fixative. Resuspend the cell pellet and leave for 10 min. Repeat this step once more. The quality of the final preparations will largely depend upon the thoroughness of fixation. Most importantly, the fixative must be made up fresh, immediately before use, the cells resuspended thoroughly before fixation, and the total time of the fixation steps must be adequate to ensure the complete dissolution of erythrocytes (a total time of at least 30 min).

(10) Spin at 100g for 10 min after the last fixation and resuspend the cells in a small volume of fixative, e.g. 0.5 ml.

(11) Have ready some microscope slides which have been cleaned thoroughly before use, if necessary by washing in acid–alcohol (99% ethanol : 1% concentrated HCl) and drying. Remove any particles of dust or tissue before use. Drop one drop of cell suspension out of a Pasteur pipette on to a cleaned microscope slide from a height of 5–8 cm. Air dry the slide by waving it gently and examine it using phase contrast optics before making any further preparations. First check that the concentration of cells is suitable; if it is too high, dilute the cell suspension with a little more fixative; if there are too few cells, spin at 100g for 5 min and resuspend in a smaller volume of fresh fixative. An ideal cell suspension is one which, on splashing, gives a spread of cells which is sufficiently thin to allow unequivocal recognition of most distinct chromosome sets, without overlapping sets or an appearance of clumping of nuclei due to an excessive cell density. However, the cell density should not be so thin that only one or two nuclei are visible in each field of view using a 16× objective. Repeat the splash preparation and recheck the cell concentration. Secondly, check that the chromosomes have spread sufficiently. If unburst nuclei containing metaphase chromosomes are visible, or if the chromosome spreads contain overlapping chromosomes, then the hypotonic treatment was probably inadequate (see Figure 2.1). If this is the case, then it will prove very difficult to obtain good splash preparations. Two ways of attempting to remedy this situation are either to drop a drop of the cell suspension from a greater height (up to 40 cm) on to the slide or to pass the slide through a bunsen flame, so allowing the fixative to ignite after splashing (see Section III for details). Both of these modifications should result in better spreading of the chromosomes. If, on the other hand, the chromosomes are too well spread, so that it is difficult to distinguish whole complements (Figure 2.1), then use a less severe method of splash preparation and drop the cell suspension from a height of only 1 cm while holding the slide at a slight angle so that excess liquid drains away. Once the

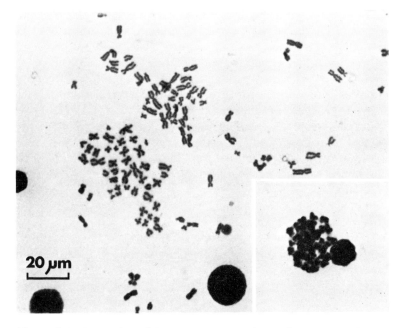

Figure 2.1 Examples of leucocyte preparations which have received excessive or (inset) insufficient hypotonic treatment. Excessive hypotonic treatment results in chromosome complements which are so disperse that they are impossible to identify as a discrete set, and which are frequently overlapping with other sets as above. Conversely, on normal splashing, a nucleus which has been insufficiently swollen by the hypotonic treatment will not burst, or if it does, the chromosomes will not separate

cell concentration is judged to be correct, prepare as many slides as you require. Any unused cells can be stored in fixative at 4°C in a closed vial, but before using such cells at a later date, spin them down and resuspend them in fresh fixative. An example of a well spread metaphase is shown in Figure 2.2.

(12) If only bright field optics are available, or if a simple staining technique is required, staining the chromosome preparations in Giemsa is recommended. Stain the slides in a 2% solution of Giemsa (Gurr's R66), in phosphate buffer ($Na_2HPO_4 : KH_2PO_4$), pH 6.8, for 10 min (see Chapter 4, Section I). After this period, flush the staining vessel out either with tap water or with distilled water. It is essential to avoid drawing the slide through the interface between the air and the Giemsa solution, as a metallic film forms at the interface and will result in a messy slide. After thorough washing to remove the excess Giemsa, dry the slide in air and examine using bright

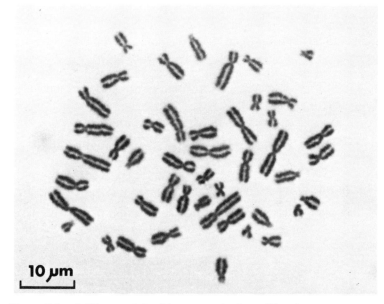

Figure 2.2   A Giemsa-stained human metaphase. The chromosomes are well separated, with no overlaps, but are maintained as a discrete set facilitating karyotype analysis

field optics and a 63× objective. More details of staining and making permanent preparations are given in Section VI.

Using the technique outlined above, reproducible results can be obtained where there are many clean, well spread metaphase chromosome preparations (Figure 2.2). For teaching purposes, students may attempt to set up cultures of their leucocytes using blood obtained from a finger prick. The finger tip should be sterilized thoroughly by rubbing well with a cotton wool pad soaked in ethanol and then punctured using a sterile lancet. An average of 0.1–0.5 ml of blood can easily be obtained by this method, sometimes more with good bleeders. The blood produced should be collected by drawing it directly from the finger into a sterile 1 ml syringe which has previously been heparinized by drawing 20 IU of heparin into it. Whole blood (0.4 ml) can then be used to inoculate culture vials containing a total of 5 ml of the medium described in step (2) above. As the equipment required is minimal and consumable materials are easily available, this can produce an easy and interesting laboratory practical exercise.

The technique that has been outlined for the preparation of human chromosomes is similar to that used in most human cytogenetics laboratories. It is a valuable diagnostic and research technique with possibilities for the detection of chromosome abnormalities and gene localization studies. The ease of setting up this technique means that it is not prohibitive in terms of cost to a research worker wishing to run through it as a 'one off' procedure. It is possible to buy chromosome kits containing all the consumable materials that are needed. The cost of these kits is prohibitive if the technique is routinely carried out, or if many samples are to be prepared, but they can be very useful for a researcher with a limited programme and no existing facilities for chromosome studies.

III. Variations on the human leucocyte technique

A. Adjustments to the preparation of slides

One of the main variations of the basic technique for preparing human chromosomes outlined above is in the method used to make the splash preparations. The simplest, and in our opinion the most reliable and the most commonly used method, is the one already described, but variations abound. The height from which the cell suspension is dropped on to the slide can be varied. Slides can be used dry, ice cold and dry, and after dipping in distilled water (without drying) – the 'wet slide' method. A major variation of the technique involves passing the slide, with the drop of cell suspension on it, through a bunsen burner flame so that the fixative ignites. The flame should not be too hot, or the slide held in it for too long, it is only necessary to ignite the fixative. Alternatively, dip the slide in methanol before dropping the cell suspension on to it and flaming, holding the slide in forceps for obvious reasons! This method of drying the slide has the advantage over air-dried preparations that the chromosomes are better spread, but this is only an advantage if, for example, the hypotonic treatment was insufficient and preparations could not be successfully made by air drying. A major disadvantage to flame drying is that the efficiency of any subsequent banding technique is reduced markedly (see Chapter 4).

B. Other mammalian species and other tissue types

The method of preparation of human chromosomes that has been outlined can be adapted to suit any mammalian species and should result in consistently good chromosome preparations. Mammals provide a convenient source of blood and in many cases venepuncture with a hypodermic needle is unnecessary. For example, blood can be obtained from a rabbit by sterilizing the ear using ethanol and nicking one of the veins in the ear using a sterile lancet or razor blade. During this procedure, the rabbit should be wrapped tightly in a sheet or towel and held firmly. Sufficient blood will be obtained from the ear to set up at least one culture. Blood samples can be obtained from a variety of rodents by snipping the tip off the tail after sterilizing with ethanol. Such sources of material are excellent for teaching purposes when the species used is not important. For many purposes, however, it is necessary to prepare chromosomes from a specific animal and occasionally this may prove troublesome. Larger and more aggressive mammals will have to be anaesthetized before a blood sample can be taken. The animal may not be a commonly available laboratory animal and the researcher may have to contact a zoo in order to obtain a blood sample. Under no circumstances will a zoo allow a valuable animal to be anaesthetized simply to take a blood sample for an outside research project. However, many routine examinations of zoo animals are carried out under anaesthetic and zoo authorities will usually oblige with a small blood sample at such a time.

Although the techniques described so far refer directly to short term culture of leucocytes or of cells growing in culture, the technique for splashing cell suspensions to obtain chromosome spreads can be adapted for most cell types. Obviously, to make splash preparations the cells must be in the form of a suspension of mainly single cells, but by enzymatic treatment most, if not all, tissues can be reduced to a single cell suspension. The versatility of such an approach is shown below in a technique for preparing chromosomes from a variety of mouse tissues.

(1) Inject the mouse intraperitoneally with colchicine. 0.1 ml of a 1 mg/ml stock solution is adequate for a good sized mouse. After 1.5 h kill the mouse. This can be accomplished either by etherization or by breaking the neck.

(2) Remove the required tissues from the mouse: suitable tissues include the testis (this will provide both spermatogonial mitoses and meiotic figures), the spleen, the intestine, and the liver. Obtain free cells by chopping the tissue into small pieces using sharp scissors, then incubating the chopped tissue in 0.1% collagenase dissolved in any minimal medium (e.g. Eagle's MEM, Earle's MEM, Hank's MEM.). Incubate for 20–30 min at 37°C with intermittent pipetting until a fine cell suspension is obtained. This can be monitored by examining an aliquot using phase contrast microscopy.

(3) Spin the cells at $100g$ for 10min; resuspend in 0.075 M KCl and spin immediately at $100g$ for 2 min; resuspend again in 0.075 M KCl. Leave for 10 min. Spin at $100g$ for 5 min, remove the supernatant, and then proceed as for step (8) onwards in the human chromosome preparation technique (Section II.B).

The above technique results in good quality splash preparations from a number of different tissues. Such an approach may be of importance if examining animals which may have a mosaic karyotype.

Another tissue which may be treated in the same way is bone marrow. Mitotic cells can be accumulated by metaphase block (for mice 0.1 ml of a

1 mg/ml stock for 1.5 h; for larger mammals a larger amount, which may have to be empirically determined) and after killing the animal the bone marrow cells can be obtained by aspiration of the long bones (femur and humerus), in larger mammals the sternum as well. The easiest way of aspirating the bone marrow cells is by using a syringe with a needle of an appropriate size (a Pasteur pipette can be used for rabbits or large mammals), with a little isotonic saline or MEM in the syringe to help flush out the cells. Treatment of the bone marrow cells subsequent to this should be as from step (6) onwards in the human chromosome technique (Section II.B).

It should be emphasized at this stage that the preparation of mitotic chromosomes from any mammalian cell culture line is essentially the same as has been described in the preceding sections. The main difference lies in the treatment of the cells prior to the hypotonic step, as most cell lines form monolayers which may need to be detached from the culture vessel using trypsin. A useful review of karyotype analysis of cell culture lines is given by Worton and Duff (1979).

C. All mitotic stages

In some cases it may be desirable to isolate chromosomes in prometaphase rather than in metaphase. In prometaphase the chromosomes are much less condensed than in metaphase and lateral differentiation of the chromosomes using certain banding techniques is therefore much easier and clearer. This is especially true for G-banding when there are many more clearly distinguishable bands in prometaphase. Treatment with spindle inhibitors not only causes an accumulation of metaphases by disruption of the spindle, but also causes additional condensation of the chromosomes. If fine detail banding studies are to be carried out it may therefore be wise to omit the colchicine treatment, but otherwise to prepare the chromosomes exactly as described. Obviously there will also be anaphase and telophase figures present, so all stages of mitosis can be examined.

The techniques for obtaining chromosome preparations from mammalian sources can also be modified to suit avian and reptilian sources. There are, however, some important modifications required in the technique and these will be demonstrated, using some specific examples, in the next section.

IV. Non-mammalian sources

Non-mammalian sources can be subdivided into two categories: warm-blooded (homeothermic) animals and cold-blooded (poikilothermic) animals. Birds belong to the first category and the differences between the techniques for obtaining chromosome preparations from mammals and birds lie mainly in the ways of obtaining material. The required temperatures for *in vitro* culturing are the same as for mammals, as are the ranges of concentration and time for spindle inhibition and hypotonic swelling. Avian leucocytes can be cultured using similar techniques to those used for mammalian leucocytes. Another common source of dividing material in birds is feather pulp material and a technique using this is described in Section V.B. A typical technique is presented here using avian leucocytes to demonstrate the minor differences from those using mammalian material.

A. Avian leucocytes

The following technique was developed by Shoffner *et al.* (1966).

(1) Take 0.5–10 ml of blood from the bird, the amount depending on the size of the bird and the number of cultures required. Blood can be taken either from a wing vein or by cardiac puncture. Neither of these methods should cause the death of the bird, although for an investigator who is unused to bleeding birds, the wing vein technique is to be recommended. The blood should be drawn into a syringe of suitable capacity and containing 0.1 ml heparin.

(2) Transfer the blood into a sterile 10 ml vial, add phytohaemagglutinin M in the ratio 0.2 ml PHA : 10 ml blood. Mix well and chill at 4°C for 30 min. Spin the tube lightly ($30g$ is adequate) for 5 min. Remove the upper plasma layer which will contain the leucocytes. An average yield will be about 1.5 ml of plasma plus leucocytes from an initial 10 ml of whole blood. Inoculate 1 ml of this plasma into a sterile culture vial containing the following: 7 ml of a suitable culture medium (e.g. Eagle's Basal MEM or Medium 199 with Earle's salts); 1.5 ml of cell-free serum (autologous, produced by centrifuging the blood from which the leucocytes have been removed at $200g$ for 5 min and removing the upper cell-free plasma); 0.15 ml of phytohaemagglutinin P (note *not* the M form, Difco cat. no. 3110-56-4); 100 IU/ml penicillin, 100 μg/ml streptomycin.

(3) Incubate at 37–38°C for 72 h. There is no need to shake the cultures during this period as the red blood cells were agglutinated and removed before culturing. For the final 1–3 h of culture time, add colchicine to a final concentration of 0.25–0.5 μg/ml of culture medium. Swirl the contents of the vial to suspend the cells and transfer to a centrifuge tube. Spin at $100g$ for 10 min and remove the supernatant. Resuspend the cells in 5 ml of 0.45% sodium citrate which has been prewarmed to the culture temperature. Incubate for 5–15 min at 37–39°C. (Some techniques recommend 0.9% sodium citrate for up to 30 min). Spin at $100g$ for 10 min and remove all but 0.5 ml of the hypotonic solution, in which the cells should be resuspended. Follow the procedure outlined for the preparation of human chromosomes from step (8) onwards (Section II.B). Wherever variable temperatures or times of

incubation are mentioned in the above protocol, the precise conditions must be determined empirically. The middle value of the ranges given should, however, give adequate preparations for most avian species.

For poikilothermic vertebrates the first factor that has to be changed for optimal culture growth conditions is temperature. For most cold-blooded vertebrates, the temperature of culturing is slightly above that of the living animal and a rough temperature guide is given in Tables 1 and 2 in Wolf (1979). Culture media will also vary, but in general a commercially available medium can be modified simply to suit most vertebrates. An indication of the necessary modifications is given in Wolf (1979). The reader is referred to specific texts on culturing poikilothermic material for further details (for reviews see Kruse and Patterson, 1973). For leucocyte culture, 20% fetal calf serum is widely used, but for some species, notably some elasmobranchs, homologous serum is preferred because of the urea concentration in the serum and the minimal amount of serum albumin present. Leucocyte cultures can be made from poikilothermic vertebrates using similar methods to those used with homeothermic vertebrates. The blood will separate on standing or mild centrifugation into two layers, the upper of which will contain the leucocytes. The leucocytes can then be cultured at a suitable temperature and treated exactly as for mammalian leucocytes. Note, however, that the concentration of the hypotonic treatment may have to be adjusted in proportion with any modifications in the culture medium. In most cases it is possible to take blood samples without killing the animal, but this will require practice and it may be best to anaesthetize the animal lightly and bleed by direct cardiac puncture. Alternatively, if the animal is to be killed anyway, phytohaemagglutinin and colchicine can be injected before killing to obviate the need for *in vitro* culturing. Such a technique will now be described for snakes, but this could be modified to suit other vertebrates.

B.   Snake leucocytes

This technique is due to Baker *et al.* (1971).

(1) Inject the snake intraperitoneally with phytohaemagglutinin. Normally PHA is supplied as a lyophilized powder which is then reconstituted to a final volume of 5 ml. For this technique, dilute one vial of PHA to 75 ml and inject 0.02 ml/g body weight for the first 35 g, then 0.01 ml/g for each additional gram. Repeat 24 h later.

(2) Forty-two hours after the first PHA injection, inject with aqueous colcemid at 0.25 mg/kg body weight, then after a further 4 h incubation bleed by cardiac puncture into a heparinized syringe and mix well.

(3) Transfer the whole blood into a vial, allow to stand at room temper-

ature for 15 min, and then centrifuge slowly for 3 min to separate the leucocytes. Remove the upper plasma layer containing the leucocytes and spin at 100*g* for 10 min.

(4) Discard the supernatant and resuspend the cells in 2–5 ml of 1% sodium citrate. Leave for about 10 min, spin at 200*g* for 3 min, and remove all but 0.5 ml of the hypotonic supernatant. Proceed as for steps (8) onwards in the human leucocyte technique (Section II.B).

C. Amphibian intestinal epithelium cells

Although there is no theoretical barrier to the preparation of amphibian chromosomes from leucocytes, it is a difficult technique to use and produces variable results. In the past most mitotic material was obtained from the gonads, but a useful technique has been developed by Stephenson *et al.* (1972) for the preparation of mitotic spreads from intestinal epithelium cells. The following version of that technique is one that gives reliable and reproducible results.

The animals should be in good condition and well fed both before and during the colchicine treatment.

(1) Inject the animal intraperitoneally with a 1.5 mg/ml (0.15%) solution of colchicine to give a final concentration of 75 μg/g body weight. Alternatively, colcemid may be used at the same final concentration. Keep the animal well fed for a maximum of 48 h (minimum of 24 h). Kill the animal by immersing it in the anaesthetic MS 222 (see Chapter 6, Section III A for full details), this will normally take about 15 min. At the end of this period wash the animal well under running water and sever the spinal cord at the base of the skull. Make a clean incision along the entire ventral surface to expose the stomach and intestines. Remove the entire intestine from the rectum to the anterior end of the stomach. Do this carefully, snipping away any attached connective tissue.

(2) Slit the intestine along the entire length; this is most easily accomplished by using a pair of fine scissors, inserting the point inside the intestine at one end and then snipping along the intestine as if cutting open a piece of tubing. This step may be carried out dry, or with the intestine immersed in the hypotonic solution (0.05 M KCl). Rinse any mucus or food debris very thoroughly out of the exposed intestine with 0.05 M KCl. Transfer to fresh 0.05 M KCl and incubate for 20–30 min. Take into account the length of time the intestine has previously been in the hypotonic solution when timing this step.

(3) While the intestine is incubating in the hypotonic saline, gently scrape the loose epithelial cells away from the intestinal wall. This can be done using the rounded end of a small spatula or a similar instrument. Do not use

a sharp instrument, such as a scalpel, or chunks of the intestine may be removed and this will subsequently hamper fixation and splashing. By the end of the period of incubation in hypotonic saline, the cells from the gut epithelium should be in suspension and can be centrifuged at 100*g* and fixed as described in steps (7) onwards of the human chromosome technique (Section II.B). An alternative to making a splash preparation could be to squash the cells and a protocol for this technique will be described later in this chapter.

D.   Amphibian bone marrow cells

This technique is a very simple method for the preparation of a large number of mitotic spreads from amphibians. It is not, however, suitable or all amphibian species, but has been tried successfully with a variety of *Rana* and *Bufo* species. We are grateful to Dr Zhengan Wu for details of this technique.

(1) Inject the animal intraperitoneally with a 0.005% colchicine solution 15–20 h before killing. In some cases higher concentrations of colchicine may be necessary. The amount of injected solution will range from 0.3 to 1 ml, depending upon the size of the animal.

(2) Kill the animal by immersion in MS222 (see previous section) or by exposure to ether vapour (Chapter 6). Rinse under running water and sever the spinal cord. Remove both femurs and scrape away as much connected tissue as possible. The tibias can also be used and if a lot of material is required, the humerus is another source.

(3) Cut off the ends of each femur to reveal the bone marrow and flush out the marrow with 0.046 M KCl. This should be carried out by filling a 1 ml plastic disposable syringe with 0.046 M KCl (the hypotonic solution) and using a needle of suitable gauge for the width of the bone marrow cavity (Figure 2.3). Insert the needle into the bone marrow cavity and gently expel the marrow by flushing some hypotonic KCl through the cavity. It may help to move the needle around in the marrow to detach it from the bone. The amount of KCl used to flush out the marrow depends on the cell density. As the marrow is flushed out, add it dropwise to clean, dry microscope slides and check the cell density using a low power phase contrast objective. It is preferable to have a thin suspension at this stage, so adjust accordingly. If the suspension is too thin, add more than one drop per slide, but if too thick, add more KCl to dilute. Normally two or three slides can be prepared from the marrow in each femur.

(4) Add two or three drops of hypotonic solution to each slide and very gently spread the cell suspension over the surface of the slide. This should be done by holding a thin glass rod (or Pasteur pipette) about 2–3 mm above the surface of the slide, touching the cell suspension with the rod, and,

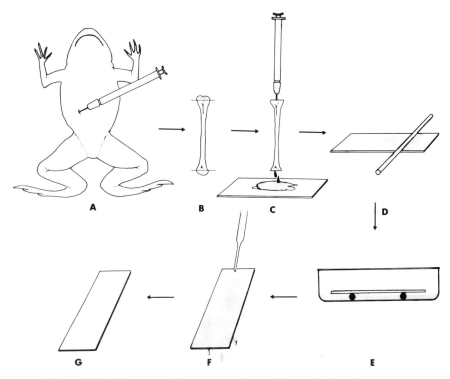

Figure 2.3 Diagrammatic representation of the technique for preparation of mitotic chromosomes from amphibian bone marrow. A, Injection
of the animal with colchicine. B, After killing the animal, remove the
femur, and clean away very thoroughly all muscle and connective tissue.
Then cut off both ends of the bone, slicing just distal to the widest part
of the bone at each end. C, Flush the cavity of the bone with 0.046 M
KCl, carefully collecting the flushed out material in a pool on a clean
microscope slide. D, Spread the cell suspension over the slide with a glass
rod. E, Fix in a moist chamber. F, Treat with 1 : 2 ethanol–glacial acetic
acid. G, Air dry. See text for fuller details

maintaining the same height, gently drawing the rod across the whole slide.
It is essential that the rod does not scrape across the surface of the slide or
the cells will be broken.

(5) Place the slides in a moist chamber and leave for 30 min at room
temperature. A suitable vessel to use to make a moist chamber is a glass
staining dish with dimensions of at least 10 cm × 10 cm × 5 cm. Into the
bottom of the container place a thick (1 cm diameter) U-shaped glass rod
and place the slides on this before covering the dish firmly. Check periodically
that the slides are not drying out. Sufficient hypotonic solution should have
been added to the cell suspension to prevent drying out during the 30 min

incubation. This obviates the need for placing any of the hypotonic solution in the bottom of the moist chamber. It is also important to have a close fitting lid on the chamber.

(6) After 30 min, place some fixative solution in the bottom of the chamber, ensuring that it does not touch the slides at all. The fixative solution is absolute ethanol – glacial acetic acid – distilled water in the ratio of 1 : 2 : 3. Leave for 2–2.5 h at room temperature, firmly covered. The cells become fixed by the vapour of the fixative solution.

(7) Remove the fixative solution and replace with absolute ethanol, again ensuring that there is no contact with the slides. Leave for 10–20 min at room temperature.

(8) Remove the slides from the moist chamber and, holding each one at an angle of 45°C, gently pipette absolute ethanol–glacial acetic acid (1 : 2) over the slide. Repeat three or four times. Allow to air dry.

(9) Stain with 2% Giemsa in phosphate buffer, pH 6.8, for 10–15 min.

V.   Mitotic chromosomes using the squash technique

Preparation of mitotic tissue from insect material is nearly always carried out using testis material where both spermatogonial mitosis and meiotic chromosome preparations are produced. One exception to this is in the Diptera, for which superb mitotic preparations can be obtained from larval neuroblasts using a method of squashing the tissue. A method for such preparations is described in the next section.

The techniques described in this section are for preparation of mitotic material from intestinal epithelium in Amphibia, from neural ganglia in Diptera, and from feather pulp in birds. The principles behind these techniques are the same as for splash preparations of tissue: colchicine arrest at metaphase, hypotonic swelling, and fixation of the material, but the tissue is then squashed as described more fully in Chapter 3.

A.   Amphibian intestinal epithelium tissue

This method does not differ at all from that given for air-dried splash preparations except in the final stages.

(1) The gut is dissected out and treated with 0.05 M KCl as described in the previous section, but the loose epithelial cells are not scraped away from the gut wall during this hypotonic treatment (Section IV.C, step (3)).

(2) After incubation in the hypotonic solution, the gut is lifted out and most of the excess liquid allowed to drain off. The whole gut is then placed in fresh, ice-cold 3 : 1 ethanol–acetic acid (or methanol–acetic acid), left for

at least 10 min, and then transferred to fresh fixative. The material can be stored in fixative at 4 °C until required for use, or it may be used immediately. If it is stored, change the fixative before using the gut and leave for a few minutes before proceeding.

(3) Trim about 5 mm off the intestine and place it on a very clean, siliconized coverslip. Add one drop of 45% acetic acid and place under a low power dissecting microscope with a black background and bright incident light. Gently scrape the loose epithelial cells away from the muscle wall of the gut, taking care not to tear the muscle. Remove any lumps of tissue, food, etc., so that a suspension of cells from the inner epithelium layer of the gut is left. Gently lower a clean, dry, subbed slide on to the cell suspension until it touches the surface of the liquid. Try to prevent air bubbles from forming when doing this. Air bubbles are easier to avoid if the cell suspension is prepared on a siliconized coverglass rather than on a slide, as surface tension maintains the liquid in a discrete droplet.

(4) Place the slide, coverglass up, in the fold of a piece of filter paper and gently blot away any excess acetic acid. Now squash firmly on the coverglass with the ball of the thumb, taking care to avoid any sideways motion of the coverglass, and examine the preparation using phase contrast optics to check that the chromosomes have spread well. Place the preparation on dry ice for 10 min or dip in liquid nitrogen for 15 s, then flick the coverglass off using a razor blade and place it immediately into a Coplin jar containing 95% ethanol. Leave for about 30 min, remove, and air dry. For full details of the dry ice technique for obtaining permanent squash preparations see Chapter 3.

Intestinal epithelial preparations can be stained using acetic–orcein instead of 45% acetic acid when squashing (see Chapters 3 and 5). Briefly, the epithelial cells are scraped away from the muscle wall in a solution of 1% orcein in 45% acetic acid and are then left for 10 min before squashing. Stained preparations can be mounted in Euparal immediately after removal from the ethanol.

B. Avian Feather Pulp

Feather pulp material is frequently preferred over leucocytes as a source of mitotic material for two reasons. First, the material is readily available and does not involve any risk to the bird; secondly, the material is actively dividing and does not require culturing, so that chromosome preparations can be made quickly. The technique that will be described is taken from Shoffner *et al.* (1966).

(1) The best type of feather to use is an immature non-pigmented feather. For this reason feathers from newly hatched specimens are ideal, but small

body feathers from adults can also be used. The feather pulp from young birds is also easy to swell and to macerate, whereas that from primary or secondary wing feathers of adults may be very tough and difficult to swell and squash. Colchicine or colcemid is used to arrest the dividing feather pulp cells at metaphase and these spindle inhibitors can be injected either into the peritoneal cavity or into the base of the feather shaft. If the latter course of action is followed, mark the injected feather with a marker pen. A 0.05% solution of colcemid is injected into the peritoneal cavity or into the wing vein 30–45 min before harvesting the feather. Forty microlitres of stock colcemid per 30 g body weight (for young birds), or 1 ml per 1.3 kg (for adults), is used. From 25 to 50 μl is injected directly into the soft sheath of the wing feather about 1–2 cm from the base. The exact amount of colcemid needs to be determined empirically, but the above figures should serve as a guide.

(2) After a 30–45 min colcemid treatment, pluck out the treated feather and remove the pulp. This is most easily accomplished by squeezing the feather shaft between the points of a pair of blunt forceps and pushing the pulp out of the proximal end of the feather by running the forceps towards that end. Squeeze the pulp directly into a hypotonic solution made up of 0.45% sodium citrate in distilled water. The pulp is semi-solid and particularly in feathers from older birds it must be macerated slightly while in the hypotonic solution. Do not, however, attempt to make a single cell suspension at this stage.

(3) After 10–20 min incubation in the hypotonic solution, transfer the small pieces of feather pulp to 45% acetic acid and leave them in this fixative for at least 30 min before proceeding.

(4) Remove a single piece of feather pulp from the fixative and remove excess liquid by touching very lightly on a piece of tissue or filter paper. Then smear the feather pulp over about 1 cm^2 of the surface of a clean subbed slide by rubbing it in a circular manner over the surface until a thin layer of material can be seen. Take care that the preparation does not dry out, if necessary add a small drop (about 5 μl) of 45% acetic acid. Remove any larger pieces of pulp or obvious clumps of cells and touch a siliconized coverglass on to the preparation. Squash as described in the previous method, except that slightly more pressure may have to be exerted to get satisfactory squashing of the cells. Slides should then be made permanent using the dry ice and ethanol method and can be stained or mounted as required.

C. Neuroblasts of dipteran larvae

Neural ganglia are the most convenient source of mitotic tissue in Diptera and the method used to prepare metaphase chromosomes parallels that used for mammalian tissue.

(1) Third instar larvae of *Drosophila* are used as a source of neural ganglia, which is convenient in that mitotic preparations from the neural ganglia and polytene preparations from the salivary glands can be made from the same larva. Third instar larvae that have left the surface of the culture medium but are still mobile are used (see also Chapter 5 for fuller description of the stages of *Drosophila* development). The neural ganglia (dorsal and ventral), are dissected out after decapitation of the larva as described in Chapter 5, Section IV.B. The neural ganglia are small and extremely fragile and should be handled as little as possible during subsequent stages. A number of dorsal ganglia can be dissected out at the same time and placed in a watch glass containing 5 ml of Schneider's embryonic cell culture medium (Schneider, 1964), at room temperature. The following technique for the preparation of metaphase chromosomes from *Drosophila* neural ganglia has been taken from Cobel-Geard and Gay (1980). This method can be adapted to many other Dipteran species.

(2) When a sufficient number of neural ganglia have been accumulated in Schneider's medium, 40 µl of a 100 µg/ml refrigerated stock solution of colcemid is added and mixed in gently; incubation is continued for 45 min at room temperature (final concentration of colcemid is 0.7 µg/ml).

(3) Transfer each ganglion to a separate clean subbed slide by picking it up in a droplet of medium held between the two points of a pair of fine forceps – do not grasp the ganglion with the forceps. Add one drop of 1% trisodium citrate ($Na_3C_6H_5O_7.2H_2O$) and incubate for 10 min. Gently remove most of the hypotonic solution using a Pasteur pipette which has been drawn to a fine point in a bunsen flame.

(4) Immediately add 10 µl of 45% acetic acid to each ganglion, taking care that none of the ganglia dry out. Leave to fix for 5 min, then gently lower a siliconized coverglass on to each preparation. Gently tap the cover-glass with the eraser end of a pencil or the wooden handle of a dissecting needle a few times, taking care not to move the coverglass horizontally. Place the slide, coverglass up, between the fold in a piece of filter paper and gently blot away any excess fixative. Apply slight pressure by rolling the index finger across the coverglass through the filter paper to spread the chromosomes. Check using phase contrast optics that the chromosomes have spread sufficiently and, if they have, make the preparation permanent by the dry ice and ethanol method.

Chromosomes obtained from larval neural ganglia can be stained in 2% Giemsa in phosphate buffer, pH 6.8, and mounted in Euparal–DPX–Permount as described in the next section.

IV. Making Permanent Preparations

For simple staining of metaphase chromosomes, a 2% solution of Giemsa in 0.01 M sodium phosphate buffer, pH 6.8, is routinely used. Slides are stained for 10 min and are washed in running tap water to avoid drawing the slide through the air/liquid interface (see also Chapter 4, Section I, for further details of Giemsa staining). One reliable type of Giemsa has been found to be Gurr's improved R66, which is used for staining all types of chromosome preparations. After staining, the slides are air dried. They can then be examined directly without mounting, or made permanent using mounting medium. The following protocol works well.

After air drying the stained slides thoroughly, dip them in a Coplin jar containing xylene. Allow them to stand in xylene for a few seconds, lift them out, and drain off the excess xylene. Do not allow them to dry out completely. Place the damp slide on a folded sheet of filter paper and put one drop of Permount over the preparation. Gently lower a clean dry coverglass on to the slide, then lightly blot away the excess xylene and mountant using the filter paper and dry on a hotplate (45 °C) overnight.

References

Baker, R. J., Bull, J. J., and Mengden, G. A. (1971). Chromosomes of *Elaphe subocularis* (Reptila: Serpentes) with the description of an *in vivo* technique for preparation of snake chromosomes. *Experientia (Basel)*, **27**, 1228–1229.

Barker, B. E. (1969). Phytohaemagglutinin and lymphocyte blastogenesis. *In Vitro*, **4**, 64–79.

Bashor, M. M. (1979). Dispersion and disruption of tissues. In *Methods in Enzymology*, Vol. LVIII, *Cell Culture* (W. B. Jakoby and I. H. Pastan, eds), Academic Press, New York and London, pp. 119–131.

Blakeslee, A. F., and Avery, A. G. (1937). Methods of inducing doubling of chromosomes in plants. *J. Hered.*, **28**, 392–411.

Cobel-Geard, S. R., and Gay, H. (1980). A new simplified method for the preparation of neuroblast mitotic chromosomes from *D. melanogaster*. *Drosophila Inf. Serv.*, **55**, 148–149.

Cunningham, B. A., Sela, B. A., Yamara, I., and Edelman, G. M. (1975). Structure and activities of lymphocyte mitogens. In *Mitogens in Immunobiology* (J. J. Oppenheim and D. L. Rosenstreich, eds), Academic Press, New York, pp. 13–30.

Dick, C., and Johns, E. W. (1968). The effect of two acetic acid fixatives on the histone content of calf thymus deoxyribonucleoprotein and calf thymus tissue. *Exp. Cell Res.*, **51**, 626–632.

Eigsti, O. H., and Dustin, P., Jr (1955). *Colchicine: In Agriculture, Medicine, Biology and Chemistry*, Iowa College Press, Ames, Iowa.

Farnes, P., Barker, B. E., Brownhill, L. E., and Fanger, H. (1964). Mitogenic activity of *Phytolacca americana* (Pokeweed). *Lancet*, **ii**, 1100–1101.

Ham, R. G., and McKeehan, W. L. (1979). Media and growth requirements. In *Methods in Enzymology*, Vol. LVIII, *Cell Culture* (W. B. Jakoby and I. H. Pastan, eds), Academic Press, London and New York, pp. 44–93.

Hsu, T. C. (1952). Mammalian chromosomes *in vitro*. I. The karyotype of man. *J. Hered.*, **43**, 172.

Hsu, T. C. (1979). *Human and Mammalian Cytogenetics. An Historical Perspective*, Springer Verlag, Berlin, Heidelberg, and New York.

Hughes, A. (1952). Some effects of abnormal tonicity on dividing cells in chick tissue cultures. *Quart. J. Microsc. Sci.*, **93**, 207–220.

Jakoby, W. B., and Pastan, I. H. (eds) (1979). *Methods in Enzymology*, Vol. LVIII, *Cell Culture*, Academic Press, London and New York.

Klášterská, I., and Ramel, C. (1979). The hypotonic pretreatment in mammalian cytology: its function and effect on the aspect of meiotic chromosomes. *Hereditas*, **90**, 21–29.

Kruse, P. F., Jr, and Patterson, M. K., Jr (1973). *Tissue Culture: Methods and Applications*, Academic Press, New York.

Levan, A. (1938). The effect of colchicine on root mitosis in *Allium*. *Heredity*, **24**, 471–486.

Nowell, P. C. (1960). Phytohaemagglutinin: an initiator of mitosis in cultures of normal human leucocytes. *Cancer Res.*, **20**, 462–466.

Schneider, I. (1964). Differentiation of larval *Drosophila* eye antennal discs *in vitro*. *J. Exp. Zool.*, **156**, 91.

Shoffner, R. N., Krishnan, A., Haiden, G. J., Bammi, R. K., and Otis, J. S. (1966). Avian chromosome methodology. *Poultry Sci.*, **XLVI**, 333–344.

Stephenson, E. M., Robinson, E. S., and Stephenson, N. S. (1972). Karyotype variation within the genus *Leiopelma* (Amphibia: Anura), *Canad. J. Genet. Cytol.*, **14**, 691–702.

Sumner, A. T., Evans, H. J., and Buckland, R. A. (1973). Mechanisms involved in the banding of chromosomes with quinacrine and Giemsa. I. The effects of fixation in methanol–acetic acid. *Exp. Cell Res.*, **81**, 214–222.

Wolf, K. (1979). Cold blooded vertebrate cell and tissue culture. In *Methods in Enzymology*, Vol. LVIII, *Cell Culture* (W. B. Jakoby and I. H. Pastan, eds), Academic Press, New York and London, pp. 466–477.

Worton, R. G., and Duff, C. (1979). Karyotyping. In *Methods in Enzymology*, Vol. LVIII, *Cell Culture* (W. B. Jakoby and I. H. Pastan, eds), Academic Press, New York and London, pp. 322–344.

Additional references not cited in text

ACT Technical Manual (M. G. Wahrenburg and Nancy Schanfield, eds.). A publication of The Association of Cytogenetic Technologists. Address for further details, M. G. Wahrenburg, Pfizer Inc. Research, Eastern Point Road, Groton, CT 06340, U.S.A.

Hsu, T. C. and Pomerat, C. M. (1953). Mammalian chromosomes *in vitro*. II. A method for spreading the chromosomes of cells in culture. *J. Hered.*, **44**, 23–29.

Priest, J. H. (1977). *Medical Cytogenetics and Cell Culture*, 2nd edn, Lea and Febiger, Philadelphia.

3

Chromosomes in meiosis

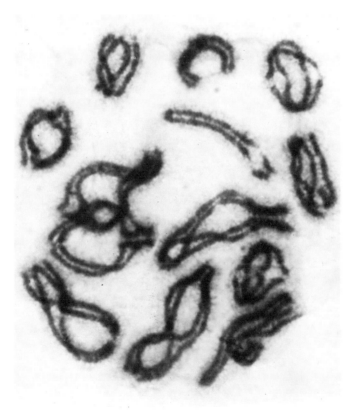

A set of diplotene chromosomes from the testis of *Oedipina uniformis*, a plethodontid salamander from Central America. The XY chromosome pair, in end to end association, can be seen just above and to the right of the centre of the picture. [The photomicrograph was kindly provided by Dr James Kezer of the University of Oregon. Reproduced with permission from Novitski, E., *Human Genetics*, Macmillan, New York (1977)]

This chapter deals with methods for the study of chromosomes in meiosis employing mainly the squash and splash techniques. Most early work on meiosis, including the classical 19th century work of Janssens (1905), as well as the vast majority of the work described by E. B. Wilson in the 1925 edition of his famous book, *The Cell in Development and Heredity*, was carried out on material that had been fixed, emedded in wax, sectioned, and stained with haematoxylin. Although the methods involved in this kind of approach will receive scant treatment here, this is not to say that wax embedding and sectioning techniques have lost their value or have been superceded by squashing and splashing. On the contrary, they are still exceedingly valuable, especially when it is necessary to examine relatively undistorted cells from various aspects, so that the distribution of chromosomes within cell nuclei and on division spindles can be appreciated. Squash and splash techniques are fast and efficient methods for looking at chromosomes, and just that. Embedding and sectioning are for looking at chromosomes and cells *in situ* and in relation to surrounding structures and cells within the same tissue environment.

Meiosis is generally represented as a formal process during which a germ cell passes through a number of quite distinct phases, each of which is characterized by peculiarities of chromosome shape or behaviour. It is usually studied in males, simply because spermatocytes are small and abundant and their chromosomes are easily accessible; eggs are, conversely, large and relatively few, and as a rule their chromosomes are hard to find and harder still to prepare for critical examination. Later in this chapter, some references will be given for those who wish to study pre- or post-diplotene events involving the chromosomes of female amphibians and insects, but for the most part what follows will be directed at those who wish to study male meiosis in animal testes.

When embarking on a study of a meiotic system, students and researchers alike should take full account of two matters, both of which will help them to understand what they see and to extract the most from their preparations. First, the process of meiosis is remarkable for its timing and its condensation and decondensation cycles. It begins with a long critical introduction, the prophase, that is highly significant for its consequences but exceedingly hard to analyse from the cytological standpoint. It ends with a burst of activity from diplotene through to second anaphase when the chromosomes are almost continuously moving and changing form. During this cycle the chromosomes at one time or another pass through almost every conceivable state of condensation and decondensation, assuming highly distinctive and supposedly functional forms at each stage. From the practical standpoint, an appreciation of the timing of meiosis is most important. Figure 3.1 shows an example of this timing as worked out by Callan and Taylor (1968) for the newt *Triturus vulgaris*. It will be helpful to remember, when examining preparations of meiotic tissues, that the longer a cell remains in a particular stage the more common will cells in that stage be in any preparation, and *vice versa*. For example, most meiotic preparations abound in cells in leptotene and pachytene, but cells in second meiotic prophase are exceedingly rare.

The second matter to bear in mind is that meiosis is an immensely variable process from one organism to another. For example, even within the salamander genus *Batrachoseps*, which includes the animal employed

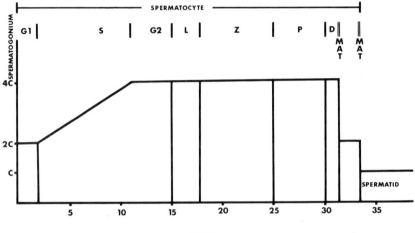

Figure 3.1 Diagram to illustrate the timing of meiosis from the end of the last spermatogonial mitosis through to the end of the second meiotic anaphase in the newt *Triturus vulgaris*. The diagram is presented as a graph of the nuclear DNA content in terms of the *C* quantity (haploid equivalent) of DNA. Spermatogonia begin with the 2*C* amount of DNA embodied in a diploid set of chromosomes each of which comprises just one chromatid. Premeiotic S-phase then produces spermatocyte nuclei with the 4*C* amount of DNA in the diploid number of chromosomes, each chromosome having two chromatids. The two meiotic divisions then produce a stepwise reduction of the nuclear DNA content to 1*C* and the chromosome content to haploid. Note the long first meiotic prophase extending from day 15 through to day 30 or beyond

by Janssens in 1905, some species have substantial amounts of centromeric heterochromatin and others have none. Some have a highly diffuse diplotene stage inserted between pachytene and diplotene, whereas in other species this stage is lacking. In some organisms the chromosomes are organized into a polarized array, often described as a bouquet, during zygotene and pachytene. Other species have no bouquet or show an entirely different form of polarization. Some species start chromosome pairing at the middles of the chromosomes, others at the ends, and others at apparently random points all along the lengths of the chromosomes. Be aware of this variability and think how it has proved helpful in exposing the really significant and purposeful features of the meiotic process.

The squash technique provides a simple and fast way of visualizing chromosomes and cell nuclei in animal and plant tissues. It is carried out in four main steps: (1) fixation of the tissue in ethanol (or methanol)–acetic acid; (2) softening and maceration of the tissue in 45% acetic acid; (3) squashing between slide and coverglass; (4) making permanent.

Fresh tissue is dissected from an animal and placed immediately in a freshly made ice-cold mixture of three parts of ethanol (or methanol) to one part of glacial acetic acid (3 : 1). It can be left in this fixative for days, even years, so long as it is kept below 4 °C. The usual practice, if not making permanent preparations immediately, is to place the fixed tissue in 3 : 1 and keep it in the freezing compartment of a refrigerator. If preparations are to be made immediately, then the tissue should be left in cold 3 : 1 for 1–2 h.

Ethanol–acetic acid preserves chromosome morphology well and it is simple and easy to use. It is the cytologists' 'favourite fixative'. It has its limitations, however, and if an experiment requires careful preservation of certain structural or chemical components of cells, then it would be better to think hard about alternative methods of preparation and fixation. Methanol–acetic acid is frequently employed in mammalian chromosome research, particularly when there is reason to suspect that stocks of ethanol may contain substantial amounts of water. The two basic rules with regard to fixation for subsequent squashing are first, use ethanol in preference to methanol unless there are good reasons for doing otherwise, and second, avoid the use of formalin or fixatives that irreversibly denature and coagulate proteins.

After fixation, the tissue is softened by immersion in 45% acetic acid. There is nothing magical about the number 45. It is just that this concentration of acetic acid is known to strike about the right compromise between inadequate softening and over-hydrolysis of the tissue. The acetic acid has the effect of softening collagenous materials and causing cells to swell and fall apart from one another. It is easy to watch how opaque, white, newly fixed material becomes transparent and gradually falls apart in 45% acetic acid. Of course, 45% acetic acid is a fairly drastic treatment for any cell, and many of the cellular proteins are lost in this step. Amongst other things, the nucleolus tends to disappear in 45% acetic acid, although most of its RNA remains *in situ*, and virtually all cytoplasmic structure is lost. The softened tissue can easily be teased apart into what is as nearly as possible a single cell suspension. This is usually done using about 1 mm^3 of tissue in a drop of 45% acetic acid on a microscope slide. When the tissue has been teased apart a coverglass is placed on top of the preparation, excess liquid is removed by placing the slide in a fold of filter paper and blotting the area round the coverglass, and the preparation is then squashed by thumb pressure on the region of the filter paper that overlies the coverglass. The preparation can then be examined with a microscope, although it is still temporary and if not sealed or made permanent it will dry out in due course and become useless.

At one time making a squash preparation permanent presented some difficulty. Cells are squashed between a slide and a coverglass and these must somehow be separated from one another in such a way that the cells remain firmly attached to the slide or the coverglass and are not moved, damaged, or lost. They must then be dehydrated in alcohol and finally mounted in a medium that will preserve them and allow them to be examined by critical microscopy. The condition of the final preparation must be such that it can be stored over a period of years and examined and cleaned repeatedly. Now clearly, it is not possible simply to take the coverglass off a newly made squash preparation and expect to proceed with satisfaction. The result is catastrophic!

In 1953 Conger and Fairchild published the answer to this problem. The method they suggested had in fact been in use for some time previously, but to them must go the credit for having published the details for the benefit of other investigators in the field. Essentially, when making a squash use a slide that is thinly coated with gelatin (a subbed slide) and a coverglass that is thinly coated with silicone (siliconized coverglass). After the squash is made, put the slide on the surface of a flat block of dry ice (solid CO_2) or immerse it in liquid air or nitrogen (the latter is usually obtainable from electron microscopy laboratories). Dry ice is generally safer, simpler, and cheaper. This effectively freezes the tissue to the sticky surface of the subbed slide. The tissue does not stick to the water-repellent surface of the siliconized coverglass. After everything has frozen hard, and this takes about 5 min, flick off the coverglass with a razor blade. The cells will remain on the slide and the coverglass will come away perfectly clean. The slide is then placed immediately in 95% ethanol and the preparation subsequently dehydrated and mounted in the normal fashion.

It is probably fair to say that the Conger and Fairchild technique revolutionized cytology, and it has since been used in virtually every study that has involved looking at chromosomes from compact animal and plant tissues. To be sure the entire squash technique is full of compromises and ways that have been adopted simply because they seem to work. Nevertheless, the beauty of the technique is its simplicity and speed, and every investigator is to be encouraged to adapt it to his or her own particular needs and introduce improvements and modifications as appropriate.

Staining of squash preparations can be carried out in any one of three principal ways. The material can be stained most easily by dissolving carmine or orcein in the 45% acetic acid that is used to soften the tissue, in which case the material is stained immediately before squashing and actually squashed in the stain solution. The material can be stained with the Feulgen reaction before placing in 45% acetic acid for softening and squashing, or the squash preparations can be dehydrated and air dried and subsequently stained with any of a range of different dyes, the most widely used being Giemsa. Protocols will be included in this chapter for staining with orcein and Feulgen. Methods for staining air-dried preparations with Giemsa are given in Chapters 2 and 4. Some useful variations in the use of orcein and certain other stains are given in Chapter 5. Although these latter methods are particularly recommended for use with polytene chromosomes, they are entirely applicable in many other situations and should be tried if it seems sensible to do so.

I.   Choice of animals

Three kinds of animal will feature in what follows: salamanders, locusts, and mice. The reasons for selecting these animals are simple. Salamanders have large genomes and relatively low chromosome numbers. They therefore have large chromosomes. At certain times of year their testes contain cells in all stages of meiosis. The animals are abundant and easily collected from the wild or obtained from suppliers in the United States, and the same can be said of newts from the standpoint of European investigators. Locusts have large chromosomes relative. to most other insects. At middle to late fifth instar their testes contain cells in most stages of meiosis. Their cytology is in many respects typical of orthopterans, a group that has been a source of many major contributions to our understanding of the meiotic process. Locusts can be reared without difficulty in the laboratory or obtained from suppliers. However, they are strictly controlled in the United States and are therefore difficult to obtain in that country. During the summer months in temperate regions or throughout the year in the tropics, grasshoppers and crickets are often locally abundant and easily caught with a light sweep net or a butterfly net. They should certainly be explored as likely sources of good meiotic material. Mice need no special advocacy other than to say that they are probably the simplest system on which to learn techniques for studying meiosis in male mammals.

A.   Salamanders

The animals covered in this section are specifically salamanders belonging to the family Plethodontidae. Most of the species mentioned in this chapter can be obtained from suppliers listed in Appendix 1, although, of course, it is easier to get precisely what is wanted at the right time of year by going out and collecting it; and plethodontid salamanders are unbelievably common and abundant in many parts of the United States. Indeed some species, particularly in the eastern United States are so common as to be caught and sold as fish bait!

All plethodontids in North America are seasonal and the testes are in the best condition for meiotic studies in the first half of the year. Species in the western United States, i.e. California, Oregon, and Washington, begin meiosis earlier than those in the east or mid-west. Western species are best collected between January, when most of the testis will consist of cells in early prophase, and May, when the testis will consist largely of stages from late pachytene through to spermatozoa. Persons planning a research or teaching programme that requires meiotic material should think ahead, make their collections at the right time of year, and if need be store the testes fixed in 3 : 1 in the freezer until required.

Plethodontids have no lungs and breathe for the most part through their skin. Accordingly they are moisture dependent and die very quickly indeed if placed in dry or hot conditions. The commoner species are to be found in moist woodland areas under rocks, logs, and other pieces of debris. They are most easily found after long periods of rain when the ground is thoroughly wetted. They are very hard to find and give the impression of being extremely scarce when the weather is hot and conditions on the surface are dry. But after a day of rain and a brief cool spell in the months of April to July in, for example, New York, New Jersey, or Connecticut, it is usual to be able to collect several hundred animals in less than an hour from a square kilometre of woodland. All the collector needs is a small pick for turning over logs and stones, some quick reactions because the animals can move quite fast, a plastic bag with some wet paper towelling in it to carry the animals, and a polystyrene (Styrafoam) box with some ice for transporting bags of animals back to the laboratory. Determine the normal adult size for the species in question and do not collect juveniles.

The animals are best kept at 15–20 °C in glass jars with moist paper towelling and fed on live *Drosophila*. Most North American plethodontids are impossible to sex, except in the case of gravid females that can sometimes be identified by the large light coloured eggs showing through the ventral abdominal wall.

Virtually any species of plethodontid salamander will prove good for meiotic studies if collected at the right time of year. However, plethodontids from the western United States are likely to be better than those from the east or mid-west simply because western species have larger genomes and therefore larger chromosomes. Perhaps the best animal of all is the Californian slender salamander, *Batrachoseps attenuatus*.

Full details of species characteristics and geographical distributions for plethodontids are given in field guides by Conant (1975) and Stebbins (1966).

With regard to the use of other tailed amphibians for meiotic studies, newts and ambystomid salamanders offer good material, but for the biologist working in the United States they are less abundant, less convenient, and generally less satisfactory than plethodontids. Newts of the genus *Triturus* are plentiful in Europe, but they are hard to obtain from the wild at the time when their testes are rich in first meiotic stages. *Triturus cristatus carnifex*, *T. marmoratus*, *T. vulgaris*, and *T. alpestris* can easily be kept in the laboratory all year round and will come into good meiotic condition if maintained in a reasonable simulated natural environment with regard to temperature, day length, and availability of food. However, this is an expensive way of getting material, particularly for teaching purposes, and teachers and researchers outside the USA may prefer to rely on more easily obtainable insect material. Comments on the availability and maintenance of European newts are given in Chapter 6. Frogs and toads are plentiful in most parts of the world and are easy sources of meiotic material; but they all have quite small chromosomes and in this sense they are definitely not good for teaching.

In summary then, teachers and researchers in North America and Canada are encouraged to explore the Plethodontidae as a rich source of superb meiotic material, but since this is only available at certain times of year forward planning is essential. Persons working in Central America or the northern parts of South America are the most fortunate of all since they have access to the tropical plethodontids that not only have very large and interesting chromosomes but are meiotically active at all times of year. Those working in Europe and other parts of the world will find it rewarding to explore the excellent cytology of tailed amphibians that are available locally and such species as can be obtained from overseas, but orthopterans will prove cheaper and more effective as teaching material.

B. Locusts and Grasshoppers

Schistocerca gregaria, the desert locust, is probably the easiest animal to rear and maintain in the laboratory for use as a source of meiotic material. Full details for rearing and breeding locusts are given in a most helpful pamphlet produced some years ago by the Anti-Locust Research Centre (Hunter-Jones, 1961). Locusts at various stages of development can usually be obtained from suppliers listed in Appendix 1.

The point in development at which a locust testis contains cells in all stages of meiosis varies from one species to another, but, as a rule, with *Schistocerca* and *Locusta migratoria* the fifth nymphal instar is likely to be best. In *Locusta*, testes from mid to late fifth instar males usually have everything from pachytene through to late spermatids, with a usefully large number of cells in diplotene and first meiotic metaphase. Earlier stages, i.e. late fourth and early fifth instar, will have more spermatogonial mitoses and meiotic prophase nuclei. When starting work on some new material or setting up for a teaching exercise it is, of course, best to make preparations and check each stage beforehand. The features of the last two nymphal stages of *Locusta* are shown in Figure 3.2.

During the warmer times of year in all parts of the world there are many species of grasshopper to be caught in the wild and most of these will provide excellent material for meiotic studies. The best approach with such animals is to collect, identify, and have a look. Local species of grasshopper can be a valuable source of material for senior high school or undergraduate project work. The cytological variability in grasshoppers from one population, race, or species to another can be exceedingly interesting and significant from the standpoint of both evolution and cytotaxonomy.

The four commonest species of grasshopper in the British Isles are *Chorthippus parallelus*, *C. bruneus*, *Omocestus viridulus*, and *Myrmeleotettix maculatus*. *C. bruneus* is a wasteland or urban species, favouring dry open stony or even sandy habitats, including areas of asphalt and concrete in the vicinity of buildings. *M. maculatus* likewise favours dry localities but is more common on open heathland or sand dunes, especially where there is a good covering of grass. *C. parallelus* and *O. viridulus* are both meadow species, favouring localities where the grass is rather lush and moist. All these species are common enough to be collected in large numbers throughout June, July, and

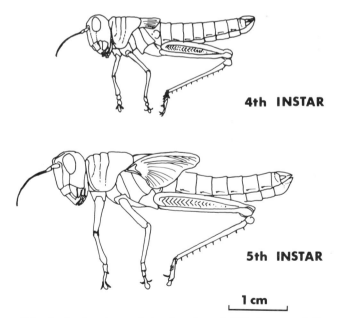

4th INSTAR

5th INSTAR

1 cm

Figure 3.2 Scale drawings of the fourth and fifth instars of *Locusta migratoria*. The most useful criteria for determining stage are overall size of the animal and the appearance of the wing pads

August. The simplest method of catching is by rapid sweeping of an area with a butterfly net or light sweep net, passing through the grass as close to the ground as possible on each sweep. More selective and sophisticated methods of catching are described in a number of entomology texts, but they are all much more time consuming and less grossly productive than vigorous indiscriminate sweeping of an area known to contain substantial numbers of grasshoppers. An excellent account of the characteristics, habitats, and distributions of British grasshoppers is given by Ragge (1965). The most common species of grasshopper in use for cytological studies in North America is *Melanoplus femur rubrum*, and several laboratories and suppliers now maintain cultures of these animals. A good source of information on the grasshoppers of North America can be found in Helfer (1953), and details of Australian species are given by Rehn (1952).

As a rule, crickets and other orthopterans are not a good source of meiotic material.

C. Mice and other mammals

For those wishing to look at meiotic chromosomes from mammals the laboratory mouse is undoubtedly the best animal to start with. Moreover, the protocol that will be given later for mouse testis works well with only minor modifications on material from most other species of mammal, including man. Unlike amphibians and orthopterans, the technique for mouse meiosis is beautifully simple and the yield of meiotic preparations will always be high. However, mouse chromosomes are small, they are numerous ($n=20$), they are all telocentric, and they are therefore a little less interesting in the cytological sense than the chromosomes of amphibians and grasshoppers. This is to some extent offset by the fact that numerous different strains of mice are available with interesting structural rearrangements: they provide opportunities for challenging cytogenetic studies. Moreover, mammalian testes are generally larger and more uniform than those of other animals and they therefore offer excellent opportunities for controlled cytochemical and cytological studies: and of course there are often veterinary or paramedical reasons for preferring mammalian material if the objective of a programme is a better understanding of meiotic systems in humans or domestic animals.

Mice belonging to common inbred laboratory strains are usually readily available from university animal houses, animal suppliers, or pet stores. Virtually any mouse will do if it is between about 2 months and 1 year of age. Wild mice are entirely satisfactory. Indeed the only small mammals that may give problems are those that are strongly seasonal in their breeding habits. Both Syrian and Chinese hamsters will give excellent meiotic preparations, and in some respects they are better than mice, especially since their chromosomes are more varied in size and shape.

It is usually possible to carry out successful testis biopsies with mammals, although this is hardest of all with small mammals where the tunica is exceedingly thin and delicate and therefore difficult to suture effectively. Testicular biopsies in the UK require a Home Office Licence for experimentation on live animals.

II.   General equipment needed

(a) A microscope fitted with a phase contrast and a bright field condenser, phase contrast 16× or 25× objectives, and 63× or 100× bright field objectives of the highest possible quality

(b) A dissecting binocular microscope with magnifications ranging from 10× to 20×

(c) Fine dissecting scissors

(d) Watchmaker's forceps no. 4 stainless steel, two pairs

(e) Subbed slides (see Chapter 9, Section V)

(f) Siliconized coverglasses, 18 or 22 mm square. These can be prepared by dipping clean coverglasses in 'Repelcote' (Hopkin and Williams cat. no. 996270, obtainable through British Drug Houses Ltd) and then standing them to dry in air, or by rubbing clean coverglasses with a siliconized tissue ('Sight-saver' tissues manufactured by the Dow Corning Corporation, USA)

(g) Plain coverglasses, 22 mm square, number 1½

(h) Whatman filter papers number 1, 12.5 cm in diameter or larger

(i) Single edge razor blades

(j) A diamond marker pencil, or another means of marking slides that will not be affected by water, acetic acid, alcohol, or xylene

(k) A pencil with a new and unused eraser at one end

(l) A 'tapper', most easily constructed by flattening and smoothing off the end of a piece of nylon or Teflon rod of 3–4 mm diameter

(m) Some clean wiping tissues or an old well washed handkerchief

(n) A camel hair brush, very carefully cleaned by dipping in ether and shaking dry, or a small can of liquid air with a gas nozzle attached (gas-jet duster or 'dust-off' compressed air can)

(o) A number of small glass vials with caps, 10–20 ml capacity

(p) Some disposable Pasteur pipettes

(q) A Petri dish filled with a 0.5% solution of MS222 or chlorotone (see Chapter 6, Section III A i), or a large glass jar with screw-on top and a wad of cotton wool soaked in 1 : 1 ether–chloroform stuck to the inside of the top (this is for use with amphibians only, and only if MS222 or chlorotone are not available)

(r) Distilled water

(s) 0.05 M KCl (for amphibians) and/or 0.075 M KCl (for mammals), for hypotonic pretreatment of tissues before fixing (see Chapter 2)

(t) Freshly made up 3 : 1, i.e. three parts absolute ethanol to one part glacial acetic acid

(u) 45% acetic acid

(v) Dry ice in a polystyrene (Styrafoam) thermal container or wide mouthed vacuum flask: dry ice is best prepared as a solid block with a perfectly flat surface of about 10 cm × 10 cm in area, or finely powdered by passing through an ice grinder and then smoothing out to form a wide flat surface that will accommodate at least five slides lying alongside one another; if dry ice is not available, then liquid air or liquid nitrogen at least 10 cm deep in a wide mouthed vacuum flask can be used; for persons who do not have access to either dry ice or liquid gases, an alternative method is given later in this chapter (see Section III.B)

(w) 70% ethanol, 95% ethanol, and 100% ethanol in quantities and histology jars or dishes suited to the number of slides to be processed

(x) Xylene

III.   Feulgen-stained squashes of amphibian testis

The structure of the testis of plethodontid salamanders has been described by Kingsbury (1902) and Burger (1937). The latter paper contains an excellent description, in part dealing specifically with *Plethodon cinereus*, and it should be consulted for detailed information. The testis of the red-backed salamander and of most other plethodontids has a cylindrical structure and consists of a longitudinal duct surrounded by ampullae that are connected by short ducts to the main longitudinal duct. Primary spermatogonia are clustered about the short ducts of the ampullae and these, along with the duct system, constitute the persistent structures of the testis. The reproductive cycle is an annual event in temperate zone plethodontids. After the ampullae have been emptied of their sperm, the testis is built up by proliferation from the persistent primary spermatogonia, so that at a particular time during the year the ampullae become filled with secondary spermatogonia that are available for transformation into spermatocytes.

The meiotic divisions appear first at the posterior end of the testis and spread through the gonad during a period of about 2–3 months in a caudocephalic wave. As a consequence it is possible to obtain salamanders in which the meiotic events are present in the testis in a sequentially ordered series; the ampullae at the extreme anterior end of the testis will contain spermatocytes in leptotene, slightly more posterior ampullae will have zygotene, and all other stages of meiosis will follow in sequential order through spermatids to sperm in the extreme posterior ampullae. It is therefore possible, as in no other species of animal, to dissect out a testis, fix it, and cut it transversely into slices to be individually squashed, each of which will have predominantly cells of one stage in the meiotic process.

The common newt, *Triturus vulgaris*, and the crested newt, *T. cristatus*, are probably the best and most readily obtainable sources of material for persons working in western and southern Europe, and both these species are available through suppliers in the UK to persons in other parts of the world. Throughout the winter and during the spring breeding season the testes of these animals consist mainly of follicles that are full of mature sperm with a small group of tiny follicles, lying dorsal and anterior, containing spermatogonia. At the end of the breeding season all sperm are voided into the vasa deferentia, the follicles that contained them degenerate, and a new round of spermatogenesis starts up in the translucent whitish part of the testis. The spermatogenic wave passes diagonally across the testis, so that part way through this process the most advanced cells lie in the ventral posterior region and the least advanced in the dorsal anterior region. A testis fixed at the appropriate time, September to December, will contain all spermatogenic stages.

The reason for recommending Feulgen staining for salamander and other amphibian testis is as follows. It is unusual to be able to collect meiotically active animals and use them soon after for cytological preparations. It is more often convenient to collect at the right time of year, fix the testes in 3 : 1, store them in the freezer, and then stain and squash them at some later date. It is often quite difficult to soften tissue that has been stored merely by immersing and macerating it in 45% acetic acid, and the longer it has been stored in 3 : 1 the more of a problem this becomes. The problem is completely overcome by the use of the Feulgen reaction. The combination of hydrolysis in HCl followed at a later stage by immersion in 45% acetic acid is guaranteed to give excellent squashes even with material that has been stored for several years. A procedure for orcein staining is given later in this chapter in the section on locust and grasshopper material. More information on the Feulgen technique is given in Chapter 10.

A. Procedure

(1) Place the animal in a Petri dish of anaesthetic (chlorotone for salamanders, MS222 for newts). If it proves difficult to anaesthetize with either of these substances, then use ether:chloroform vapour.

(2) When the animal has stopped moving and is completely insensitive, wash it copiously under the cold tap and then nick its spine with a pair of scissors just behind the head. Wash the animal again, place it on its back on a piece of moist paper towelling, open the abdomen, and remove the testes.

(3) Place the testes in 0.05 M KCl for 10 min. With larger testes (more than 5mm × 2 mm) the testis should be split cleanly down its long axis with a new razor blade before being immersed in KCl to allow better access of the saline to the inner parts of the testis. Each investigator should use

hypotonic treatment (0.05 M saline) with discretion. Some meiotic cells fix and squash very well indeed without any hypotonic treatment. With other materials hypotonic treatment is absolutely essential to obtain well spread metaphases.

(4) Place the testis in freshly prepared, ice-cold 3 : 1 ethanol–acetic acid for 1 h, or for longer periods at temperatures below 5 °C if it is to be stored in this condition.

(5) While it is fixing in 3 : 1 place the testis on a slide and cut it into pieces of 8–10 mm³, placing each piece in a separate fixing vial and labelling it with regard to its original position in the whole testis. This is particularly important when dealing with the testes of plethodontids which have meiotic stages distributed sequentially along their lengths.

(6) Remove the 3 : 1 from the fixing vial using a Pasteur pipette and taking care not to break up or damage the pieces of tissue. *Do not allow the tissue to dry out at any stage.*

(7) Then proceed with the following steps, each time removing the liquid from the vial with a pipette and pouring in the next material. All steps are carried out at room temperature unless otherwise stated.

  (i) 70% ethanol, 5 min.
  (ii) 50% ethanol, 5 min.
  (iii) Distilled water, 5 min.
  (iv) 5 N HCl, 20°C for 20 min; or 1 N HCl, 60°C for 6 min.
  (v) Distilled water, three washes of 2 min each.
  (vi) Schiff's reagent, 20°C, 1½ h. (To make Schiff's reagent add 0.5 g basic fuchsin powder to 100 ml water at room temperature, add 1 g potassium or sodium metabisulphite and 10 ml of 1 M HCl. Shake at intervals until straw coloured (1–3 h), add 0.5 g activated charcoal, shake, filter, and store in a tightly stoppered bottle in the refrigerator. The reagent should be water clear and completely colourless).
  (vii) SO_2 rinses, two of 10 min each (SO_2 rinse consists of 10 ml 1 M HCl, 10 ml 5% potassium or sodium metabisulphite, 180 ml water).
  (viii) Distilled water.

(8) Clean off a series of subbed slides by lightly wiping them with fine, lint-free tissue or a handkerchief. Clean off a series of siliconized coverglasses. The meticulous cleaning of slides and coverglasses at this stage is absolutely crucial for success. Any hard foreign material left on slide or coverglass will prevent effective squashing of cells and chromosomes. Remember, a chromosome may only be 0.5–2 μm thick. If it happens to be lying beside a bit of hair, grit, or cellulose fibre that is 20 μm thick, then the chromosome will certainly not be squashed. The objective for a successful squash preparation is fastidiously clean subbed slides, fastidiously clean sili-

conized coverglasses, and nothing but lots of loose cells on the slide in a small drop of 45% acetic acid. So proceed as follows.

(9) At this stage there is a choice between a method in which the tissue is placed on the subbed slide, macerated, a siliconized coverglass placed on top, and the cells squashed between slide and coverglass, or another method in which the tissue is macerated on the coverglass and a subbed slide lowered on to the coverglass. Both ways are equally effective in terms of squashing cells to visualize chromosomes, but the latter method of macerating cells on the coverglass does seem to ensure better against loss of cells and chromosomes from the preparation during subsequent steps in making permanent. The only real disadvantage of the coverglass method is that the coverglass sometimes breaks during the maceration process. The method described below is the one in which the tissue is placed on the slide.

Put a small piece of testis on a slide in two or three drops of 45% acetic acid. Pieces of about 1–2 mm^3 are about right.

(10) Working with a dissecting binocular microscope and two pairs of very clean fine watchmakers' forceps, mince the tissue thoroughly. This means pulling it apart into the smallest possible fragments with the objective of reducing it as nearly as possible to a single cell suspension. Another method of mincing involves the use of a nylon or Teflon 'tapper'. Start with a piece of tissue in a few drops of 45% acetic acid and then, holding the tapper quite vertical, tap rapidly and quite positively on the surface of the slide in the region of the tissue fragment with the object of pulverizing the tissue into a single cell suspension. This tapping method may be too drastic for some materials. It works well with newt testis and with some non-meiotic materials such as intestinal epithelium or plant material. It is severely damaging to salamander testis. Different tissues are likely to require quite different degrees of macerating and often quite radically different approaches to the operation. The best method can only be determined by trial and error and individual inventiveness.

(11) When the tissue has been fully macerated, pick away all pieces of tough connective tissue and any other fragments of material that may be around. Do this under a dissecting microscope so as to see what needs taking away and what is left.

(12) Still working with the dissecting microscope and the tips of fine forceps, distribute the cellular material evenly over an area of about 1 cm^2.

It is important at this stage to have just the right amount of 45% acetic acid on the slide. If there is too much then most of the tissue will be scattered to the edge of the coverglass when it is placed on the preparation immediately before squashing, and it will therefore be lost during the blotting and squashing steps. If there is too little 45% acetic acid then the preparation may be in danger of drying out. In general, it is best to keep on the slide just enough acetic acid to allow good maceration and to avoid drying out. Between 20 and 30 μl is about right.

(13) Take a siliconized coverglass and give it one final blow to remove the last few specks of dust and grit and then drop it squarely on to the pool of macerated tissue.

(14) Fold a filter paper in half and place the slide, preparation side uppermost, in the fold. Very carefully blot away excess fluid and then gently but firmly apply thumb pressure to the coverglass through the overlying filter paper. Be especially careful not to 'squidge' the coverglass sideways. Immediately after squashing examine the slide and see that there are no air spaces between slide and coverglass. If there are bubbles or air spaces then they almost certainly signify the presence of grit, fibres, or other unsquashable material, and the preparation is unlikely to be successful. Proceed with it by all means, but don't hold out much hope of it being a good one.

(15) Look at the preparation with a phase contrast microscope. Remember that it is only temporary and will soon dry out, so look quickly. If it seems to be drying out then add a very small drop – about 10 μl – of 45% acetic acid to one side of the coverglass. The purpose of this preliminary examination is to see how much squashing pressure is going to be needed with the particular tissue to achieve good separation of chromosomes. Remember, it is essential to use a well adjusted phase contrast microscope, otherwise it will not be possible to see anything. If separation is not good, then place the slide back in the folded filter paper and squash again, only harder this time. With salamander material it is usually necessary to apply very heavy pressure to obtain good chromosome spreads. So don't be afraid to squash hard if need be. On the other hand, some materials need no more than the weight of the coverglass and some gentle blotting with the filter paper. If a particular batch of material is unusually difficult to squash in the sense that the cells are well preserved and spread but individual chromosome groups will not disaggregate no matter how much thumb pressure is applied, then even more effective squashing can be accomplished with the eraser end of a pencil. Place the slide in the fold of a filter paper and locate the position of the coverglass. Hold the slide and filter paper firmly in position with the first two fingers of one hand, one finger along each end of the coverglass. With the other hand hold a pencil eraser-end down and perfectly vertical above the region of the coverglass and apply a series of hard even pushes down on the coverglass, tracking back and forth to cover systematically the whole area of the coverglass. This is a particularly heavy method of squashing, but it is sometimes effective in getting just that last bit of chromosome separation and spreading that cannot be accomplished by any other means.

When the right squash technique for the material in use has been determined, proceed with the dry ice treatment and subsequent steps towards permanent preparations.

(16) Place the preparation on dry ice for 10 min or longer. It is best to use a covered dry ice container to avoid condensation and ice forming on the surface of the slide.

(17) Remove the preparation from the dry ice and *immediately* breathe very quickly and warmly on the coverglass – just a quick 'huff' of warm air to slightly thaw the material from the inner surface of the coverglass – and then instantly flick off the coverglass with a razor blade and place the slide in a Coplin jar of 95% ethanol. The whole of this operation from removal of the slide from dry ice to dunking in ethanol should take less than 5 s.

(18) Leave the slide in 95% ethanol for 15 min. Then proceed through two changes of 100% ethanol, xylene, and mount in Canada balsam or other suitable medium.

Note that if the preparations have not been stained with Feulgen or any other dye and are intended for subsequent experimental treatments such as *in situ* hybridization or autoradiography then they can be air dried at this stage directly from the 95% ethanol. If the preparations are to be used for quantitative Feulgen microdensitometry, then they are best mounted in immersion oil after treatment with xylene (see Chapter 10).

B. Permanent preparations with no freezing

The following method for separating slide from coverglass after squashing does not require dry ice or nitrogen. It only needs patience. However, it does have the distinct advantage that it gives squashes in which the chromosomes and nuclei are much better preserved that they would be after the Conger and Fairchild procedure.

Make a squash preparation in the usual way, and immediately apply a very small quantity (about 5 μl) of 10% glycerol in 45% acetic acid to one edge of the coverglass. The coverglass must not rise as a result of adding the glycerol–acetic solution. As the medium surrounding the cells evaporates, glycerol is drawn under the coverglass. Add more glycerol–acetic every few hours, and then leave the preparation overnight in the refrigerator. Provided that sufficient glycerol has penetrated under the coverglass the preparation will not dry out. Continue adding glycerol–acetic little by little over a period of up to 2 or 3 days until glycerol has fully replaced the medium in which the cells were originally squashed. Then invert the preparation in a dish of 95% ethanol. The coverglass should quickly detach, and the flattened cells will be found adhering to the slide.

We are indebted to Professor H. G. Callan, FRS, for providing us with this method.

C. Looking at squashes of amphibian testis

The most effective equipment for examining squash preparations of urodele testis where the cells and chromosomes are large and well stained is a compound binocular microscope fitted with a 10× or 16× objective for scanning purposes and a 40× (NA 1.0) or 63× (NA 1.4) bright field objective for detailed observation and photomicrography. A green filter is useful for enhancing contrast.

A good preparation will show a high density of cells over much of the area of the coverglass with large patches of nuclei all in the same stage of meiosis. Usually, pachytene and spermatid nuclei will be most abundant and large areas of cells in first meiotic metaphase with nearby cells in late diplotene will be especially conspicuous on account of the dense staining of metaphase chromosomes. The kinds of cells that are present in the preparation will depend on the region of the testis from which the preparation was made.

Meiotic prophase nuclei of plethodontid salamanders are especially rewarding in the sense that they usually show distinct polarization of the chromosomes with all the ends directed towards one side of the nucleus. The individual chromosome threads are clearly distinguishable and have quite different appearances and thicknesses before and after synapsis. Chromocentres, formed by the clumping of centromeric heterochromatin are common. Diplotene and prometaphase are particularly handsome on account of the large size of the chromosomes and the clarity with which it is possible to see centromeres, chiasmata, and the four-strand structure of the bivalents. Suffice to say here that plethodontid salamanders are material *par excellence* for meiotic studies and those who have the opportunity to study them should sieze it. Good representative pictures of meiosis and descriptions of the process in the genus *Plethodon* are to be found in papers by Kezer (1970) and Kezer and Macgregor (1971).

Table 3.1

Species	Haploid chromosome no., n	C value (pg)
Batrachoseps attenuatus	13	42
Bolitoglossa subpalmata	13	56
Notophthalmus viridescens	11	35
Plethodon cinereus	14	20
Plethodon dunni	14	39
Salamandra salamandra	12	32
Triturus cristatus	12	23
Triturus vulgaris	12	24
Taricha granulosa	11	30

The chromosome numbers and C values (amount of DNA per haploid chromosome set) for some common urodeles are given in Table 3.1. The chromosomes of most urodeles are metacentric or submetacentric, with a few notable exceptions (Macgregor and Jones, 1977; Kezer and Sessions, 1979). Certain remarkable variations and cytological situations are worth looking out for. The striking differences in *C* values amongst the Plethodontidae are particularly significant, the animals with higher *C* values having chromosomes that are truly enormous by any standards. Indeed the Central American genus *Bolitoglossa* may well be said to have the largest chromosomes in the world! Some discussion of the significance of these large genomes and of *C* value variation within genera and species is given by Macgregor (1978, 1982). The absence of chiasmata in the long arm of the longest chromosome in *Triturus cristatus* is also worth watching for and is of particular significance in relation to the development and evolution of this and related species of newt (Morgan, 1978; Macgregor and Horner, 1980).

IV. Orcein-stained squashes of locust or grasshopper testis

The structure of the testis in locusts and grasshoppers is quite uniform from one species to another, the only obvious variable being the number of follicles. A good description of the form and development of the testis in locusts is given by Nelsen (1931).

The testis in *Locusta migratoria* is a flattened fan-like array of short, closely packed tubules each leading into one of two sperm ducts that lead directly backwards to the region of the seminal vesicles, the accessory glands, and the copulatory chamber (Figure 3.3). In a late fifth instar animal the testis is 5–7 mm long. It is semi-transparent and whitish in colour. It lies closely against the dorsal body wall about mid-way along the abdomen, and it is usually densely surrounded by yellow fat body.

Each testis tubule is about 3 mm long and shaped like a dumbell, being thicker at the distal end. The arrangement of cells within the tubule is simple. Spermatogonial and early meiotic prophase nuclei are at the distal end of the tubule and meiosis is progressively more advanced through to spermatids or mature sperm at the proximal end where the tubule joins the spermatic duct. The cells within the tubules are arranged in cysts. There are about 64 spermatocytes in each cyst and the cyst is surrounded by interstitial cells and membrane. The cells of the cysts develop synchronously. The timing of meiosis has been worked out for *L. migratoria* by Moens (1970). At 37°C premeiotic DNA replication (S-phase) lasts about 24 h. A further 24 h passes between the end of S-phase and the beginning of leptotene. Preleptotene to metaphase 1 occupies 4–5 days. Metaphase 1 to spermatid lasts 1–2 days. The timing of meiosis in *Schistocerca* is not markedly different but it is known to be quite temperature dependent, the whole process lasting 14 days at 30°C and only 6 days at 40°C.

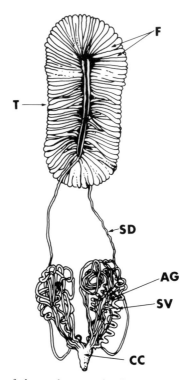

Figure 3.3   Diagram of the male reproductive system of *Locusta migratoria*. T = testis, F = follicles, SD = sperm duct, AG = accessory glands, SV = seminal vesicle, CC = copulatory chamber

A. Procedure

The locust testis is easy to find. In the method given here the testis is removed together with fat body and then teased away from the fat body in insect saline prior to fixation in 3 : 1 ethanol–acetic acid.

(1) Select a male by examining the lateral and ventral aspects of the tip of the abdomen, if need be with a hand lens or binocular dissecting microscope. The male is much simpler, ending in a single stout scoop-like sternum that sweeps upwards at the tip to form a blunt point (Figure 3.4).

(2) Remove the head, legs, wing pads, and the tip of the abdomen from the locust. Pin the animal out on its front in a dissecting dish. Insert a pair of fine scissors into the rear end of the abdomen and make two laterodorsal slits, one on each side, along the whole length of the abdomen. Dissect free the remaining dorsal strip of integument over the whole length of the abdomen, taking care not to disturb the underlying fat body and testis. The testis now lies exposed and easily identifiable on top of a mass of fat body. Remove

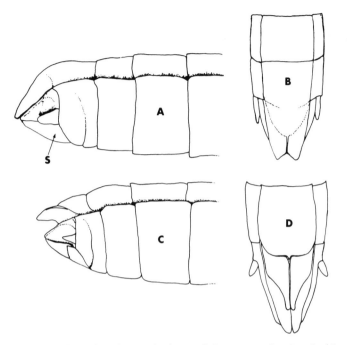

Figure 3.4   Side and underneath views of the rear ends of male (A and B) and female (C and D) *Locusta migratoria*. The male appears much simpler, ending on the ventral surface with a single stout scoop-like sternum (S)

the whole testis with as little of the fat body as possible and place it in an embryo cup of clean insect saline. Tease the testis free from all traces of fat body. Then transfer it directly to fresh ice-cold 3 : 1 ethanol–acetic acid.

For a beginner, the identification of a late fifth instar male and dissection of its testis free from fat body into 3 : 1 is likely to take as long as 15 min. The experienced investigator can normally accomplish the whole procedure in less than 2 min.

A quick procedure that is useful for grasshoppers, which are usually smaller than locusts, is to snip off the tip of the abdomen and then simply squeeze the abdominal contents out into insect saline, like toothpaste from a tube. The testis can then be sorted out from other debris under the dissecting binocular.

Testes can be stored in 3 : 1 in a freezer for use later in the year, or they can be fixed for a minimum of 1 h. Stored material that has been in fixative for a long time will be a little harder to squash. The very best squashes can be obtained from material that has been fixed for about 12 h. If material has been stored in fixative for a long time, then change it to fresh fixative before proceeding with squashing.

(3) Prepare several very clean subbed slides and siliconized coverglasses, with special attention to removing pieces of fibre, dust, hair, or any other particles that are likely to interfere with squashing.

(4) Place a whole testis in an embryo cup of 3 : 1 and pick off a number of individual testis tubules.

(5) Place four or five tubules a few millimetres apart from one another in the middle of a subbed slide. Quickly remove excess fixative with a small piece of filter paper and then cover the tubules with a few drops of lacto-proprionic orcein (4 g orcein in 100 ml of a 1 : 1 mixture of 45% lactic acid and 45% proprionic acid). Note that if preparations are required for subsequent experimental treatment and are not to be stained at this stage then the orcein can simply be omitted and the lacto-proprionic mixture used on its own. *It is most important at this stage that the material should not dry out.*

(6) Using the tapper, tap the surface of the slide rapidly and positively in the region of the tubules with the object of pulverizing the tissue into a single cell suspension. Orthopteran material usually needs vigorous and quite hard tapping for a minute or more to separate the cells adequately for good squashing.

(7) After tapping, examine the preparation under a dissecting microscope using a white background. Remove any large pieces of debris and spread the tissue evenly over about 1 cm².

(8) Take a siliconized coverglass and give it one final puff to remove dust, then drop it squarely on to the pool of macerated tissue.

(9) Move the slide very briefly through a low bunsen or spirit flame so as to warm the preparation slightly: too brief and this will have no effect; too long and the slide will crack or the preparation will cook. Warming is just about right when it begins to produce evaporation of the excess dye surrounding the coverglass. The purpose of warming is to enhance staining. It should not be used when preparations are being squashed in lacto-proprionic without orcein and are intended for subsequent drying and use in other experimental procedures such as *in situ* hybridization or Giemsa banding.

(10) Now proceed as described from step (14) onwards in the protocol for salamander material given earlier in this chapter.

B. Looking at squashes of locust or grasshopper testis

Orcein-stained preparations usually have good contrast and are best examined with a good quality compound microscope fitted with bright field objectives. Orthopteran chromosomes are generally smaller than those of urodeles so the microscope should be equipped with 63× and 100× objectives as well as a lower power objective for scanning. If for some reason the orcein staining is not sufficient to give good contrast, then it sometimes helps to use phase contrast for examining and photographing either the freshly made squash or the final mounted preparation, whilst being appropriately cautious with image interpretation.

Good preparations from late fifth instar or young adults of *Locusta* will show a wide mixture of stages right through from pachytene to later spermatids. Figure 3.5 shows the kind of distribution of cell nuclei and chromosome sets that may be expected in a good squash from a grasshopper or locust testis. In tubules from younger animals gonia and leptotene nuclei will be present and spermatids will be absent. It is not possible to distinguish leptotene from zygotene nuclei. Pachytene nuclei are usually common and characteristically show the X chromosome as a conspicuous heterochromatic mass that is quite distinct from the other chromosome material. The X chromosome in orthopterans should not be confused with the chromocentre of salamander spermatocytes at a corresponding stage. Diplotene and early first meiotic metaphase are undoubtedly the most useful and informative stages in orthopterans, and they should be abundant in animals of the right age. If cells in the same stage of meiosis lie close to one another in groups

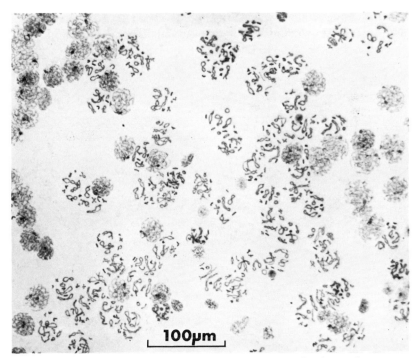

Figure 3.5 A low power view of a typical orcein-stained squash preparation of testis from the grasshopper *Chorthippus parallelus*. Nuclei in pachytene occupy the upper left corner and middle right of the picture. Most of the remainder of the field is occupied by chromosomes in diplotene

that represent the contents of the original cysts, and the chromosomes of each cell are well spread, not overlapping with one another and not mixing with chromosomes from neighbouring cells, then squashing has been just right and the preparation is probably about as good as it can be.

The karyotypes of most locusts and grasshoppers seem remarkably alike. However, beneath this superficial similarity there are some exceedingly interesting variations and differences and the cytology student can derive considerable pleasure from observing these differences and investigating them.

Both *Schistocerca* and *Locusta* have $n=23$ (11 pairs of autosomes and one X chromosome in the male). A typical set of diplotene chromosomes from *Schistocerca gregaria* is shown in Figure 3.6. The chromosomes are distinguishable on the basis of length, and are normally classified into long (L), medium (M), and short (S) groups, having four, five, and three chromosomes in them respectively. The X chromosome belongs to the L group. All the chromosomes in locusts are telocentric. In *Schistocerca* there are substantial blocks of heterochromatin around each centromere. *Locusta* has no such pericentric heterochromatin. The chromosomes of *Schistocerca* can readily be banded, using either the ASG or the $Ba(OH)_2 : 2\times$ SSC techniques (see Chapter 4). There is intensive banding at the centric regions of all chromo-

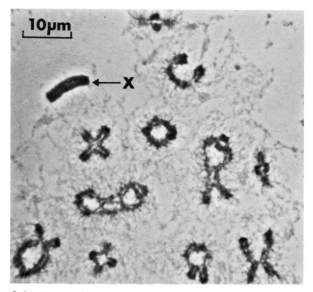

Figure 3.6   An orcein-stained squash preparation of a set of diplotene chromosomes from the locust *Schistocerca gregaria*. The highly condensed X chromosome (this animal is X0 in the male) is indicated by an arrow. The four-strand structure is clearly visible in some of the larger bivalents

somes, two chromosomes have bands at the positions of the nucleolus organizers, and the three S chromosomes have wide bands at their non-centromeric ends. The centromeric bands are the same size for all chromosomes irrespective of chromosome size, which contrasts with the situation found in salamanders where the larger chromosomes have more centromeric heterochromatin than the smaller ones.

The karyotypes of four common species of grasshopper and five species of locust are given by John and Hewitt (1966). Excellent representative pictures of pachytene, diplotene, and metaphase chromosome sets from *Chorthippus* are given by John and Lewis (1968). The Giemsa banding patterns of *Schistocerca* chromosomes are given by Fox *et al*. (1973). A method for obtaining meiotic preparations from grasshopper eggs and a striking demonstration of the differences in chiasma distributions between male and female grasshoppers (*Stethophyma*) is given by Perry and Jones (1974). A range of good general information as well as a collection of useful references is to be found in John and Lewis' (1968) article entitled 'The chromosome complement'. In addition to these selected references, the major source of information on the chromosomes and cytogenetics of grasshoppers and locusts is Michael White's excellent book *Animal Cytology and Evolution* (White, 1973).

V. Air-dried meiotic chromosomes from mice

The mammalian testis is a compact organ most of which is made up of a number of compartments or lobules that are divided from one another by septae. Each lobule encloses from one to four highly convoluted seminiferous tubules. These tubules are usually between 150 and 250 μm in diameter and may reach up to 100 cm in length. The combined length of the seminiferous tubules in all the lobules of a human testis is estimated to be in the region of 250 m. The testis tubules are usually in the form of convoluted loops, but they may branch or end blindly. At the apex of each lobule its tubules pass directly into a converging system of ducts that lead through the rete testis to the main ductus deferens (Figure 3.7).

The seminiferous epithelium of a tubule shows cells in a range of stages in spermatogenesis with the earliest spermatogonia against the outer wall or basement lamina of the tubule and the latest stages, spermatids and spermatozoa, towards the central lumen of the tubule. Spermatogenic cells are interspersed with supporting or Sertoli cells.

A section through any part of a testis lobule from a mammal will cut through the same or different tubules at several different points because the tubules are so extensively folded within the lobule. A series of spermatogenic waves pass from end to end along each testis tubule. Accordingly, different parts of a tubule show different combinations of cell types representing different phases of spermatogenesis (see Bloom and Fawcett, 1968). In small

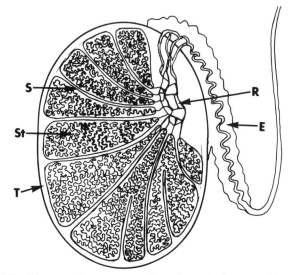

Figure 3.7   Diagramatic representation of a typical mammalian testis. T = tunica albuginea, St = seminiferous tubules, S = septa separating testis lobules, R = rete testis, E = epididymis

mammals the spermatogenic wave in one tubule has a flat front such that the range and distribution of cell types is the same all round the circumference of the tubule at any point along its length. In man, the situation is complicated by the fact that three or four different segments of the seminiferous epithelium of a single tubule are out of phase with one another with respect to spermatogenesis. Accordingly, the spermatogenic wave in man does not have the same flat front that we find in other mammals and the general histological situation appears much less orderly.

From the practical standpoint the mammalian testis is a much more homogeneous organ than that of amphibians or orthopterans. Therefore, tissue samples taken from almost any part of the testis other than the rete testis will show approximately the same ranges and proportionate numbers of cell types. This, of course, greatly simplifies matters and can be extremely valuable in the sense of providing many identical tissue samples for controlled experimentation on material from a single testis.

The timing of meiosis and spermiogenesis has been well worked out for the mouse (Ghosal and Mukherjee, 1971; Oakberg, 1978), and is as follows: premeiotic S-phase, 29–31 h; leptotene, 3 days; zygotene, 1 day; pachytene, 7–9 days; diplotene, 21 h; diakinesis, metaphase 1, and metaphase 2, 10 h; spermiogenesis, 15–16 days. The golden hamster differs only with respect to pachytene which lasts 10 days and premeiotic S-phase which lasts 40 h. It is therefore to be expected that pachytene and spermatid nuclei will be common throughout the length of the testis tubule, which indeed they are.

A. Procedure

The procedure for making splash or air-dried preparations of mouse or hamster testis is pleasingly simple and almost guaranteed to be successful. An excellent protocol for hamster material has been published by Breckon (1982), and what follows is largely based on advice kindly provided by that author. In principle, the testis is removed, its seminiferous region is opened, and the tissue tapped into a cell suspension in isotonic or slightly hypotonic medium. The cell suspension is then treated with hypotonic medium, fixed, splashed, and air dried. Preparations are then stained in one step with a combined mountant and stain. A detailed protocol for the mouse is as follows:

(1) Kill a male mouse by swift cervical dislocation or by placing it in a jar saturated with ether vapour. Do not use chloroform as this may affect the cells and chromosomes.

(2) Remove the testis complete and place it in a small dish of 2.3% sodium citrate. This should be freshly made up as it becomes quite acid after standing for a few hours at room temperature.

(3) Open the tunica of the testis and remove as much of the tissue as is needed. Avoid the rete testis as this will contribute nothing but unwanted spermatozoa.

(4) Transfer the tissue to a 5 cm diameter Petri dish half filled with 2.3% citrate.

(5) Incline the dish and macerate the testis tissue by tapping and teasing it at the edge of the pool of citrate in the dish.

(6) Remove as much of the connective tissue as possible as well as any large chunks of unmacerated material, leaving only the finest possible suspension of cells in the Petri dish.

(7) Transfer the cell suspension to one or two glass calibrated conical centrifuge tubes, and add 2.3% citrate until the volume of the liquid in each tube is about 3.5 ml.

(8) Leave standing at room temperature for a few minutes to allow the larger fragments and clumps of cells to settle to the bottom of the tube, then remove these carefully with a Pasteur pipette, leaving a fine suspension of cells in the supernatant.

(9) Readjust the volume of the 2.3% citrate to 3.5 ml and then gradually add an equal volume of distilled water, mix gently, and leave standing at room temperature for 15 min.

(10) Spin with gentle acceleration to 800 rev/min in a bench-top centrifuge for 5 min.

(11) Carefully remove the supernatant with a Pasteur pipette. The cells

are loosely packed so the supernatant cannot simply be poured off. Leave a small amount of fluid, 200–500 μl, in the tube.

(12) Flick the cells back into suspension in the remaining fluid.

(13) Gradually add freshly prepared fixative (3 : 1 ethanol (or methanol)–acetic acid) keeping the cells agitated by flicking the tube while slowly adding the fixative up to a final volume of 8–10 ml. The preparation is now being treated in the same general way as a suspension of lymphocytes from a blood culture (see Chapter 2, Section II.B).

(14) Leave for 15 min.

(15) Spin at 800 rev/min for 5 min. Remove supernatant and resuspend the cells in fresh fixative.

(16) Leave for 20 min to 2 h.

(17) Make preparations at intervals of 20 min by spinning, removing supernatant, resuspending in 2 ml of fresh fixative, and making test slides by the air-drying technique until good spreading of metaphases is observed.

(18) To make air-dried preparations take about 200 μl of the cell suspension in a Pasteur pipette. Allow two or three evenly spaced drops to fall from a height of 6–8 cm on to a clean, grease-free, warm, dry slide. Wait until the fluid from the drop has spread to its maximum extent and then hasten the drying by holding the slide in the warmth of a bench lamp and blowing gently on the surface. Examine the slide under a phase contrast microscope with a low power objective to check the density of distribution of cells. In the ideal preparation there should be about 100 clearly visible cells per field of view with a 10× objective, and of these at least one in every 50 should be in diakinesis or metaphase. If the cell density is too low, then respin, remove supernatant, and resuspend in less fixative. If the cell density is too high, then simply add more fixative.

(19) Stain with Giemsa as specified in Chapter 4, Section I, or with a combined toluidine blue mounting medium as described by Breckon and Evans (1969), the details of which are given below.

The combined toluidine blue and mountant stain of Breckon and Evans has the special advantage of being fast and exceptionally easy to use. It is ideal for teaching purposes when time is short and many students are involved. The quality of staining is not quite as good as with Feulgen, orcein, or Giemsa, but it is entirely adequate for most purposes. The stain consists of an aqueous solution of toluidine blue mixed in suitable concentration with a saturated solution of water-soluble resin. The resin is dimethylhydantoin formaldehyde (DHMF resin), obtainable as 'Brevan's mountant' from ASCO Laboratories in the UK. It has the special advantage that viscosity changes very little with increase in concentration. The mixture containing the dye can be kept fluid for months in a covered jar such as those normally used for mounting media. The medium hardens slowly on exposure to air.

To make up a solution of the resin the solid material is first broken into small pieces with a hammer and then dissolved in water little by little until about 70 g of resin have been dissolved in 30 ml of water. This may take up to a week, after which the solution, which is quite watery in consistency, is filtered and a 2% aqueous solution of toluidine blue is added until trial shows that a satisfactory staining intensity can be achieved.

Staining and mounting are carried out simultaneously by simply placing one drop of resin stain over the preparation, covering with a coverglass, and carefully blotting away excess resin with a filter paper. The preparation is immediately ready for examination. Preparations left in the light fade quite rapidly, but if kept in the dark they will retain their colour for several weeks. Faded preparations can easily be restained by gently washing off the coverglass with water and remounting in fresh resin stain.

B. Looking at mouse chromosomes

The chromosomes of mice are small. Persons accustomed to working with

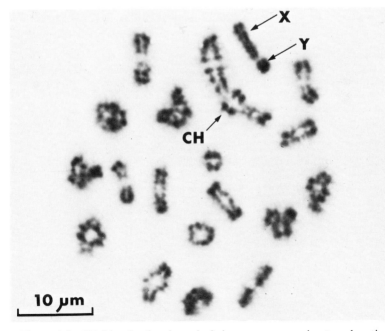

Figure 3.8   Diakinesis of male meiosis in a mouse carrying translocation T-27. The translocation is presented in this photomicrograph as a chain (CH) of four chromosomes (two pairs). The X and Y chromosomes are seen in end to end association at the top right of the picture. [This photograph was kindly provided by C. V. Beechey, G. Breckon, and A. H. Cawood, Medical Research Council Radiobiology Unit at Harwell]

other mammalian material will find this no surprise, but those who have worked with urodele material may be quite disappointed. The average chromosome length at diakinesis is between 5 and 7 μm, which in practical terms means that in order to study mouse chromosomes critically it is essential to use a 63× or 100× oil-immersion objective. For *Mus musculus*, $2n = 40$, with 38 autosomes and an X and Y. All the chromosomes have terminal centromeres. The range of length of the chromosomes in the whole set is not wide. The largest chromosome (no. 1) represents just over 7% of the genome, the smallest (no. 20) just under 3%. Only the smallest chromosome and the Y are immediately recognizable in unbanded mitotic preparations. The Y is about the same size as chromosome 20 but normally stains more intensely.

The most useful stage for cytogenetic studies is diakinesis or late diplotene (Figure 3.8). Most autosomal bivalents have one or two chiasmata. The X and Y are usually associated end to end by one chiasma, the Y being the smaller member of the pair.

References

Bloom, W., and Fawcett, D. W. (1968). *A Textbook of Histology*, W. B. Saunders, Philadelphia, London, and Toronto.

Breckon, G. (1982). A modified hypotonic treatment for increasing the frequency and quality of meiotic metaphases from spermatocytes of the Syrian hamster. *Stain Technol.*, *57*, 349–354.

Breckon, G., and Evans, E. P. (1969). A combined toluidine blue stain and mounting medium. In *Comparative Mammalian Cytogenetics* (K. Benirschke, ed), Springer Verlag, Berlin, Heidelberg, and New York, pp. 465–466.

Burger, J. W. (1937). The relation of germ cell degeneration to modifications of the testicular structure of plethodontid salamanders. *J. Morph.*, **60**, 459–487.

Callan, H. G., and Taylor, J. H. (1968). A radioautographic study of the time course of male meiosis in the newt *Triturus vulgaris*. *J. Cell Sci.*, **3**, 615–626.

Conant, R. A. (1975). *A Field Guide to Reptiles and Amphibians of Eastern/Central North America*, Houghton Mifflin, Boston.

Conger, A. D., and Fairchild, L. M. (1953). A quick-freeze method for making smear slides permanent. *Stain Technol.*, **28**, 281–283.

Fox, D. P., Carter, K. C., and Hewitt, G. M. (1973). Giemsa banding and chiasma distribution in the desert locust. *Heredity*, **31**, 272–276.

Ghosal, S. K., and Mukherjee, B. B. (1971). The chronology of DNA synthesis, meiosis, and spermiogenesis in the male mouse and golden hamster. *Canad. J. Genet. Cytol.*, **13**, 672–682.

Helfer, J. R. (1953). *How to Know – The Grasshoppers, Cockroaches and Their Allies*, Wm. C. Brown Co. Publishers, Dubuque, Iowa.

Hewitt, G. M., and John, B. (1968). Parallel polymorphism for supernumerary segments in *Chorthippus parallelus* (Zetterstedt). I. British populations. *Chromosoma (Berl.)*, **25**, 319–342.

Hunter-Jones, P. (1961). *Rearing and Breeding Locusts in the Laboratory*, Anti-Locust Research Centre, London.

Janssens, F. A. (1905). Evolution des auxocytes males du *Batrachoseps attenuatus*. *Cellule*, **22**, 377–345.

John, B., and Hewitt, G. M. (1966). Karyotype stability and DNA variability in the Acrididae. *Chromosoma (Berl.)*, **20**, 155–172.

John, B., and Lewis, K. R. (1968). The chromosome complement. *Protoplasmatologia*, **VI**, 1–196.

Kezer, J. (1970). Observations on salamander spermatocyte chromosomes during the first meiotic division. *Drosophila Inf. Serv.*, **45**, 194–200.

Kezer, J., and Macgregor, H. C. (1971). A fresh look at meiosis and centromeric heterochromatin in the red-backed salamander, *Plethodon cinereus cinereus* (Green). *Chromosoma (Berl.)*, **33**, 146–166.

Kezer, J., and Sessions, S. K. (1979). Chromosome variation in the plethodontid salamander *Aneides ferreus*. *Chromosoma (Berl.)*, **71**, 65–80.

Kingsbury, B. F. (1902). The spermatogenesis of *Desmognathus fusca*. *Amer. J. Anat.*, **1**, 97–135.

Macgregor, H. C. (1978). Some trends in the evolution of very large chromosomes. *Phil. Trans. Roy. Soc. Lond. B*, **283**, 309–318.

Macgregor, H. C. (1982). Big chromosomes and speciation amongst Amphibia. In *Genome Evolution* (G. A. Dover and R. B. Flavell, eds), Academic Press, London and New York, pp. 325–341.

Macgregor, H. C., and Horner, H. A. (1980). Heteromorphism for chromosome 1, a requirement for normal development in crested newts. *Chromosoma (Berl.)*, **76**, 111–122.

Macgregor, H. C., and Jones, C. (1977). Chromosomes, DNA sequences and evolution in salamanders of the genus Aneides. *Chromosoma (Berl.)*, **63**, 1–9.

Moens, P. B. (1970). Pre-meiotic DNA synthesis and the time of chromosome pairing in *Locusta migratoria*, *Proc. Natl. Acad. Sci.*, **66**, 94–98.

Morgan, G. T. (1978). Absence of chiasmata from the heteromorphic region of chromosome 1 during spermatogenesis in *Triturus cristatus carnifex*. *Chromosoma (Berl.)*, **66**, 269–280.

Nelsen, O. E. (1931). Life cycle, sex differentiation, and testis development in *Melanoplus differentialis* (Acrididae, Orthoptera). *J. Morph. Physiol.*, **51**, 467–525.

Oakberg, E. F. (1978). Response of spermatogonia of the mouse to hycanthone: a comparison with the effect of gamma rays. In *Physiology and Genetics of Reproduction* (E. M. Cactintio and F. Fuchs, eds), Plenum Press, New York, pp. 197–207.

Perry, P. E., and Jones, G. H. (1974). Male and female meiosis in grasshoppers. I. *Stethophyma grossum. Chromosoma (Berl.)*, **47**, 227–236.

Ragge, D. R. (1965). *Grasshoppers, Crickets and Cockroaches of the British Isles*, Frederick Warne, London and New York.

Rehn, J. A. G. (1952). *The Grasshoppers and Locusts (Acridoidea) of Australia*, Vols I–III, CSIRO, Canberra.

Stebbins, R. C. (1966). *A Field Guide to Western Reptiles and Amphibians*, Houghton Mifflin, Boston.

White, M. J. D. (1973). *Animal Cytology and Evolution*, 3rd edn, Cambridge University Press, London and New York.

Wilson, E. B. (1925). *The Cell in Development and Heredity*, Macmillan, London and New York.

4

Chromosome banding

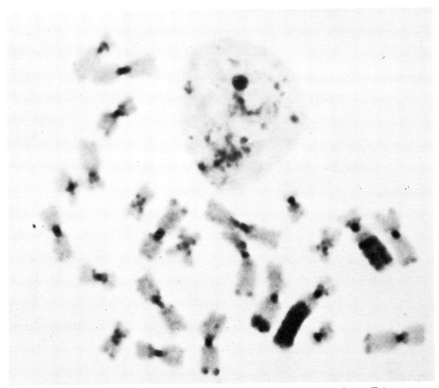

A Giemsa C-banded metaphase chromosome complement from *Triturus marmoratus* (Amphibia, Urodela). The centromeres of all chromosomes stain darkly, and especially prominent are the long arms of the largest pair of chromosomes (arrows). Magnification 1820×. [This photomicrograph was kindly provided by Simon Sims of The Department of Zoology, University of Leicester, Leicester U.K.]

Undoubtedly the necessity to develop simple and reliable techniques for distinguishing between the human chromosomes provided a considerable impetus for the development of banding methods and many banding techniques were developed specifically for human systems. However, most if not all of the banding protocols are directly applicable to material from a wide range of animal and plant species.

The first banding technique was described in 1968 and was the first technique for distinguishing between chromosomes on the basis of linear differentiation of the chromosome arms into bands. In 1968, Caspersson *et al.* described a method for banding chromosomes using a fluorescent alkylating agent, quinacrine mustard (QM), and the same group subsequently used both QM and other related compounds on chromosomes from a variety of species, including man (Caspersson *et al.*, 1969a, b, 1970). Many other techniques have now been described to allow identification of chromosomes by their banding patterns, some of the most important being C-banding (Section II), G-banding (Section III), and R-banding (Section V).

One of the most important applications of banding techniques is the unequivocal identification of the chromosomes in a karyotype. In the human karyotype, staining with Giemsa or other stains will only allow identification of a few chromosomes because of the even staining of all the chromosomes and because many of the chromosomes are similar in size and shape, rendering positive identification impossible. Similarly, in the mouse, all the chromosomes are acrocentric and form a continuous gradient in size, making identification without banding impossible.

The power of banding techniques in identifying not only whole chromosomes but also small pieces of chromosome which have undergone rearrangements is most obvious in the field of human cytogenetics. Many translocations are reciprocal and cannot be detected using simple staining techniques, whereas others may be non-reciprocal, but may involve such small amounts of material that they go unnoticed. Most, if not all, such translocations have serious clinical consequences, one of the best known examples being the Philadelphia chromosome which is implicated in chronic myelogenous leukemia (Rowley, 1973). This chromosome was originally identified as a deleted chromosome 21, but upon examination using banding techniques it was seen to be a non-reciprocal translocation of part of the long arm of chromosome 22 on to another chromosome, most usually the long arm of chromosome 9.

Assignment of linkage groups in the mouse was accomplished by using a variety of mouse strains carrying defined translocations and analysing the segregation of the linkage groups (Miller and Miller, 1972). This type of study is difficult in man where mating cannot be controlled, the generation time is long, and the family size is small. One way in which gene localization can be achieved is by taking advantage of the technique of somatic cell hybridization. Obviously a reliable method of chromosome identification is extremely important in such studies. Frequently the chromosomes of one parental species are preferentially lost, and in addition there may be chromosomal rearrangements that involve both parental chromosome sets. When rodent cells are hybridized with human cells, the hybrid cells preferentially lose human chromosomes (Weiss and Green, 1967). The presence of certain human chromosomes in clonal cell lines can then be correlated with the presence or absence of certain polypeptides that are the products of the human genes. There are several prerequisites to this type of approach. First, a reliable technique must be available to allow unequivocal identification of the chromosomes in the cell lines, not only of the human chromosomes, but also of the rodent chromosomes so that translocations can be detected should they arise. Secondly, a reliable technique must be available to distinguish the human gene product(s) from those of the rodent. Thirdly, the human gene product must be produced by the cell line under investigation. Great care must be taken to ensure that the absence of a gene product is not the result of the gene under investigation ceasing to be expressed in the hybrid cell. The last problem can be circumvented by the use of techniques to detect the gene itself rather than its product, thus eliminating the potential error presented when a gene ceases to be expressed.

The evolution of the karyotypes of many species has been studied using banding techniques and in many cases the karyotypes of related species are remarkably similar. As a direct result of banding studies, karyotype changes have been shown to arise by several mechanisms. Robertsonian fusion (or fission) can result in a reduction (or increase) in the chromosome number between closely related species. Such an event has occurred during the evolution of man and the great apes. The chromosome number of man is 46 and that of the great apes is 48. The difference lies in one chromosome pair which has undergone fusion to produce the decreased chromosome number in man (or *vice versa* if the event was one of fission). Other changes in the karyotype can occur as a result of pericentric inversions and by the addition of constitutive heterochromatin (e.g. Evans *et al.*, 1973; Pathak *et al.*, 1973; Dutrillaux *et al.*, 1975; Stock, 1976).

Notwithstanding the constancy of patterns of bands produced with any one technique and the obvious similarity between banding patterns produced by different techniques, very little is known about the kinds of DNA sequence or chromosome organization that stain with the various methods. However, the study of the mechanisms of chromosome banding is likely to help substantially towards explaining some aspects of chromosome organization. In this chapter the most commonly used banding techniques will be described, together with details of the mechanisms of banding in so far as they are understood and comments on the kinds of materials that are being differentiated. Techniques that are in common use will be described and references will be given for useful variations. In addition, details are given of two related techniques, silver staining of nucleolus organizer regions and the bromodeoxyuridine–Hoechst–Giemsa technique for visualization of sister chromatid exchange. Both are widely applicable in research and clinical studies and the silver technique at least is feasible as a laboratory teaching exercise.

I. Giemsa staining

Giemsa is perhaps the most commonly used chromosomal stain, although the precise mechanism of Giemsa staining is unclear. The active components of Giemsa are probably one molecule of eosin Y and two molecules of methylene blue which, when bound to DNA, form a magenta compound. Giemsa has no specificity for any particular base in DNA. In normal circumstances it stains chromosomes uniformly.

Two of the most reliable Giemsa stains are Gurr's R66 (Hopkin and Williams), which is supplied as a solution in methanol, and Eastman Kodak Giemsa C8685 (Eastman Kodak Co., Eastman Organic Chemicals, Rochester, N.Y. 14650, USA). Eastman's Giemsa is available as a powder (0.5 g powder mixed with 33 ml glycerol at 60°C for 2 h, then add 33 ml methanol). The protocol for Giemsa staining is very simple, but there are some important points.

(1) Dilute the stock Giemsa stain to 2% using either commercially available buffer tablets (Gurr, pH 6.8, one tablet in 1 l H_2O), or by using phosphate buffer, pH 6.8 (0.05 M phosphate buffer, pH 6.8, made by mixing 31.3 ml 0.5 M KH_2PO_4 with 22.8 ml 0.5 M Na_2HPO_4 and diluting to 500 ml). (Some banding protocols require different concentrations of Giemsa and staining times – these will be mentioned in the text as appropriate.)

(2) Stain the slide in the above stain for 10 min in a Coplin jar.

(3) After staining gently pour distilled water into the Coplin jar to flush out the stain. Dilute Giemsa tends to form a metallic scum on the surface of the dye solution and it is essential that the slide is not drawn through that or the scum will stick to the preparation.

(4) Rinse several times in distilled water for a total of 1–2 min and air dry.

(5) Slides can be examined directly after staining using immersion oil if necessary. Immersion oil can then be easily removed by dipping the slide in xylene for 2–5 min, then either gently blotting dry using a folded sheet of filter paper, or dipping briefly in acetone before air drying. If the preparation

is to be mounted, place the slide in xylene for a few moments, remove, and allow to drain slightly, then apply one drop of Permount or other xylene-based mountant with a Pasteur pipette, drop a coverglass on top, and place the slide between a folded sheet of filter paper. Gently blot away the excess xylene and mounting medium and place on a hot plate at 40–45°C to dry for 24 h.

If required, the coverglass can be removed by soaking the slide in xylene for several hours and the preparation can be destained by placing the slide in 95% alcohol for 5 min, rinsing in 70% alcohol, and air drying.

II.   Giemsa C-banding

C-banding of chromosomes was first described as a result of the pretreatment of preparations for *in situ* hybridization (see Chapter 8). These studies were carried out on mouse chromosomes during experiments to localize a mouse satellite DNA sequence. The *in situ* hybridization studies localized the satellite sequence to the centromeres of all the chromosomes (Pardue and Gall, 1970) and the authors noted that in some stained autoradiographs the distribution of silver grains corresponded to more darkly staining regions on the chromosomes. Subsequently Arrighi and Hsu (1974) modified the *in situ* hybridization pretreatment steps followed by Giemsa staining and produced a pattern of transverse bands over the centromeres, corresponding to the satellite DNA regions. In addition they demonstrated that the same technique gave a pattern of pericentric bands in other mammalian species, including man. The bands were termed C-bands because of their correlation with the centromeric heterochromatin and C-bands became associated with highly repeated DNA sequences in constitutive heterochromatin. It was known that the critical step in the production of C-bands was the denaturation of the chromosomal DNA with sodium hydroxide and this, coupled with the early observations that heterochromatin was differentially stained using this procedure, led to the assumption that the banding patterns produced were a result of differential denaturation and reannealing of blocks of DNA in the chromosomes. It was assumed that because of the greater degree of condensation of heterochromatin in the chromosome that the DNA was more resistant to denaturation and so stained more darkly with Giemsa.

C-banding procedures do in fact extract DNA from non-C-banded regions of the chromosomes (the lightly staining regions) and C-banding can be produced by mild deoxyribonuclease digestion (Alfi *et al.*, 1973). However, there is no firm correlation between the degree of condensation of a block of DNA and its ability to C-band. There is also no correlation between the type of sequence and C-banding. C-banding does stain predominantly satellite DNA-rich regions, but not exclusively so. The classical example of satellite DNA-rich C-bands are those in the mouse, where the C-bands correlate both in size and in quantity with the amount of satellite DNA. In other instances, however, C-bands do not contain appreciable amounts of satellite DNA, the best known example being the sex chromosomes of the Chinese hamster (*Cricetulus griseus*; Arrighi *et al.*, 1974). In some species, C-bands may contain more than one satellite sequence, possibly interspersed with euchromatic DNA, as in man and the primates (Gosden *et al.*, 1975, 1977).

C-bands are constant within any one species, although the size of the C-bands can be polymorphic within a species. An example of such polymorphism is the Y chromosome in man, in which the long arm has a prominent C-band that varies greatly in size. The polymorphic variants range from a virtually absent C-band to one which results in a considerable overall lengthening of the chromosome (Geraedts *et al.*, 1975).

In view of the diversity in both the types of DNA sequence present in C-bands and in the base composition of these sequences, it seems likely that C-banding results from a specific DNA–protein interaction rather than as a function of the DNA itself, as was initially postulated. However, as yet no differences in the DNA–protein composition of the chromosomes have been identified that could explain the differential C-staining, and the nature of the interaction between DNA and protein and its correlation with C-banding remains to be characterized.

A. C-banding after barium hydroxide

The procedure described below was developed by Sumner (1972).

(1) Make chromosome preparations by air drying fixed cells (Chapter 2) or by squashing the tissue in 45% acetic acid (Chapter 3). Slides should not be flame dried. The quality of the differential staining is generally improved if slides are 'aged' before being used. Ageing involves keeping the slides for a minimum of 3–4 days before use. They can be stored in a dust-free place either at room temperature or at 4°C and should be kept dry. It helps to place a small vial of dessicant in the slide box if the air is humid or if the slides are to be refrigerated. The importance of ageing slides in this way seems to be in thorough evaporation of fixative and complete drying.

(2) Incubate the slide in 0.2 M HCl for 1 h at room temperature. This step is not essential and can be omitted with very good preparations. However, if there is a lot of cell debris on the slide, recognizable as a grainy background when using phase contrast optics to examine the slide, there will be a lot of protein present and this will result in inefficient C-banding. The consistency of the technique is improved with inclusion of HCl treatment. Surplus protein is removed and the preparations suffer no damage.

(3) Rinse the slide in distilled water.

(4) Place the slide in a freshly prepared solution of 5% barium hydroxide octahydrate at 50°C for 5–15 min. A 5% solution of barium hydroxide is effectively saturated and a heavy scum forms on the surface when the solution has been standing for a while. This scum must be removed before drawing a slide through it and the easiest way to accomplish this is to draw a piece of filter paper across the surface of the solution.

(5) Rinse the slide thoroughly in several changes of distilled water.

(6) Incubate the slide in 2× SSC (0.3 M NaCl; 0.03 M trisodium citrate, pH 7.0) for 1 h at 60°C.

(7) Rinse briefly in distilled water.

(8) Stain the slide in 2% Giemsa in phosphate buffer for 1½ h. (For details of Giemsa staining see this chapter, Section I.)

(9) Rinse the slide in distilled water, dry thoroughly, and mount.

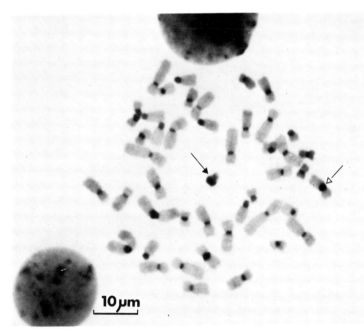

Figure 4.1 C-banded human metaphase chromosome complement. The centromeric regions of all the chromosomes are darkly stained (open arrow), as is the long arm of the Y chromosome (solid arrow). [Reproduced with permission from Sumner, A. T., *Exp. Cell Res.*, **75**, 304–306 (1972). © 1972 Academic Press Inc.]

The most critical step in this technique is the barium hydroxide treatment, which replaces the earlier and less dependable sodium hydroxide treatment.

If the incubation in barium hydroxide is too long, the chromosomes will appear swollen and distorted, an appearance that has been described as 'empty', with just the periphery of the swollen chromosomes appearing stained. If the barium hydroxide treatment is too short, the chromosomes will either be uniformly stained or the C-bands may be extremely faint or poorly defined. The most useful procedure is first to try a 10 min incubation at 50°C in barium hydroxide and then vary the time according to the degree and quality of banding. It may in some cases be necessary to reduce the temperature of incubation to 37°C and to increase the time of incubation to 1 h or more.

A long staining period in Giemsa is also essential for good C-banding and even the 1½ h time period recommended may not result in very clear differentiation. If C-bands are not sufficiently crisp and clear, then they can be accentuated by using a green filter when examining the preparation.

B.   C-banding after sodium hydroxide

This technique is due to Arrighi and Hsu (1974).

(1) Slides should be prepared by splashing and air drying. Preparations which have been made by squashing or by flame drying may also be used, but tend to yield less well banded chromosomes.

(2) Incubate the slide in 0.2 M HCl at room temperature for 15 min.

(3) Rinse with distilled water.

(4) Incubate in 0.07 M NaOH for 2 min.

(5) Rinse in 70% ethanol for 5–10 min, then in three changes of 95% ethanol over a total of 5–10 min, then air dry.

(6) Incubate the slide in 2× SSC at 60–65°C overnight. Arrighi and Hsu (1974) do not recommend incubation in Coplin jars as this leads to overstaining of the slides. Instead they recommend the following procedure. Prepare a moist chamber by placing a piece of filter paper in the bottom of a 10 cm square Petri dish. Soak the paper in a little 2× SSC, place a U-shaped glass rod on the filter paper and place the slide on this so that it is not in contact with the paper. Cover the surface of the slide with 2× SSC, place the lid on the Petri dish, and incubate as required.

(7) Rinse the slide in three 5 min changes in each of the following in this order: 2× SSC, 70% ethanol, and 95% ethanol. Air dry.

(8) Stain in 2% Giemsa in phosphate buffer for 5–30 min (see Section I).

As with the previous method, the treatment in alkali is the critical factor. Under- and overstaining result in chromosomes which have the same appearance as described for the barium hydroxide method. If C-banding is not achieved using a 2 min incubation in 0.07 M NaOH, either the time of incubation or the concentration of the NaOH can be adjusted. In the original paper, Arrighi and Hsu recommended the use of either 2× SSC or 6× SSC throughout the procedure, except at step (7) where washing in 2× SSC is suggested. If C-banding cannot be obtained using 2× SSC throughout, 6× SSC may prove more successful.

III. Giemsa G-banding

Protocols for G-banding involve pretreatment of the chromosomes that cause their structure to collapse, followed by subsequent staining during which certain regions of the chromosome reconstitute to produce darkly staining G-bands. Two basic techniques for G-banding are currently used. The pretreatment is either incubation in a saline solution or mild treatment with a protease, both of which produce the necessary collapse of the chromosome structure. Prolonged staining in Giemsa subsequent to the pretreatment allows the G-band material to reconstitute and stain. Two important prerequisites for successful G-banding are that the chromosomes are fixed in a fixative containing acetic acid and that air-dried preparations are made rather than flame-dried ones. Preparations which have been fixed in formaldehyde, for example, show uniform staining of the entire chromosome rather than lateral differentiation into light and dark bands (Stockert and Lisanti, 1972). This observation suggests that fixation in acetic acid removes material from lightly staining G-bands and fixation in formaldehyde prevents such removal and so inhibits G-banding. It is, however, not until the preparations are incubated further (in SSC or protease) that the differentiation into light and dark bands can be distinguished.

Fixation in acetic acid does cause extraction of proteins from the chromosomes (Sumner *et al.*, 1973), but to date the only demonstrable difference between light and dark bands is that positive G-bands are relatively rich in protein disulphides, whereas negative G-bands have relatively more sulphydryls (Sumner, 1974).

G-banding protocols also cause some loss of DNA from the chromosomes (Comings *et al.*, 1973), although Sumner and Evans (1973) demonstrated that there is no relationship between the amount of DNA remaining on the chromosomes and the banding pattern. There is evidence that repetitive sequences are localized predominantly in G-positive bands (Pierpont and Yunis, 1977; Yunis *et al.*, 1977), whereas sequences coding for cytoplasmic messenger RNA are localized predominantly in G-negative bands. Such observations could explain an overall difference in protein composition between light and dark bands, but such a correlation has yet to be clarified.

Positive G-bands may correspond to late replicating regions of the chromosomes, as determined both by autoradiography (Ganner and Evans, 1971; Calderon and Schnedl, 1973) and by the more accurate technique of incorporation of 5-bromodeoxyuridine (BrdU), followed by staining with Hoechst 33258 (Crossen *et al.*, 1975; Dutrillaux, 1975; Epplen *et al.*, 1975; Kim *et al.*, 1975; Latt, 1975). It is therefore possible that G-bands could correspond to regional areas of replicon organization, although a single G-band must represent many individual replicons and it is unclear how the units of replicon organization could be arranged into only two types, i.e. those represented by positive or negative G-bands.

In all situations so far examined, darkly staining G-bands correspond closely to brightly fluorescing Q-bands. Q-banding is discussed in the next section, but it is important to note here that production of Q-bands involves direct incubation of chromosome preparations with the dye without any intervening treatment, the dye binding directly to the DNA. The production of G-bands, however, is dependent upon treatment of the preparations after fixation to remove some component of the chromosome, thus allowing the Giemsa stain to bind to those regions corresponding to bright Q-bands.

The production of G-bands therefore lies in some variation of the DNA–protein (non-histone) interaction between large blocks of the chromosome which may be coordinately organized in some way as yet unknown, possibly as units of replicon organization or of transcription.

A. Trypsin G-banding

The procedure described below is due to Seabright (1971 and personal communication).

(1) Chromosome preparations should be made after hypotonic swelling and fixation in 3 : 1 methanol–acetic acid by air drying (see Chapter 2). Preparations should not be flame dried, or prepared by the wet slide method (Chapter 2, Section III.A). After the initial drop of cell suspension has spread out on the slide and the edges of the drop have begun to dry, flood the slide with two or three drops of fixative and gently drain off before allowing the slide to air dry.

(2) Leave the preparations to age in dry conditions at room temperature for 7–10 days. Slides can be artificially aged by flooding them in a 10% solution of hydrogen peroxide for 3 min, rinsing in physiological saline, and air drying. Physiological saline is 0.9% sodium chloride in distilled water.

(3) Locate a well spread chromosome set using a 40× phase contrast objective and use this set to monitor the changes during subsequent trypsin treatment.

(4) Prepare a solution of trypsin (Difco Bacto-trypsin, cat. no. 0153-60-2) by reconstituting the contents of one vial with 10 ml of sterile distilled water. This stock solution should be kept at 4°C. The working solution is prepared fresh each day by taking 1 ml of stock and diluting with 14 ml of physiological

saline. Pour a few drops of the working solution on to the slide, or dip the slide into a Petri dish or Coplin jar filled with trypsin. Leave for 5 s, then quickly wash with physiological saline and examine under phase contrast while still wet: unbanded chromosomes appear a uniform grey under phase contrast, whereas banded ones have a dark outline to the chromatids and have a 'hollow' appearance. Overbanded chromosomes appear very similar to the latter, but tend to have a very fuzzy outline. It is important that the chromosome preparations are not allowed to dry out after the trypsin treatment until they have been stained.

(5) If the chromosomes appear to have banded, stain the slide immediately in one part of Leishman stain (BDH) plus two parts of phosphate buffer, pH 6.8, for 3–8 min. The staining time may vary with the batch of stain and overstaining will obliterate some of the lighter bands. Rinse quickly in buffered water and blot dry gently using a folded sheet of filter paper. Take care not to rub the slide.

(6) If the chromosomes do not appear to have banded when viewed with phase contrast optics, repeat the trypsin treatment until they do and then stain.

(7) If after staining the chromosomes are not banded, destain by rinsing in 3 : 1 methanol–acetic acid for a few seconds and repeat the trypsin treatment.

The length of trypsin treatment required for good banding varies considerably, but generally is around 10 s. As with most staining techniques it is always preferable to under- rather than over-digest with trypsin. Giemsa can also be used as a stain, in which case the slides are stained in 10% Giemsa in phosphate buffer, pH 6.8, for 4–8 minutes, rinsed in phosphate buffer and distilled water, and air dried.

B. Acetic acid–Saline–Giemsa (ASG) banding

This method was developed by Sumner *et al.* (1971).

It is important that both the 2× SSC and the Giemsa staining solutions used in this protocol are prepared fresh each day.

(1) Chromosome preparations should be fixed in 3 : 1 methanol–acetic acid and the slides should be air dried.

(2) Incubate the slides for 1 h or more in 2× SSC (0.3 M NaCl, 0.03 M trisodium citrate, pH 7.0) at 60°C. As with the trypsin treatment, the length of incubation in SSC may need to be varied with different batches of slides. It is preferable to increase the time of incubation (up to 18 h if necessary), rather than increase the temperature.

(3) Rinse the slides briefly (for a few seconds) in distilled water.

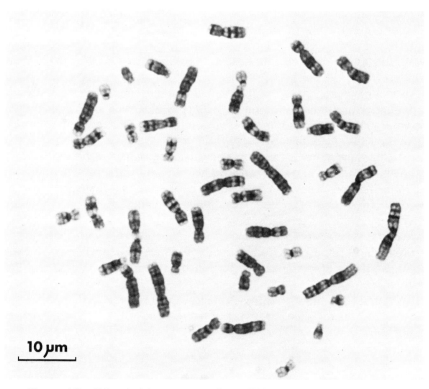

10 µm

Figure 4.2   G-banded human metaphase. This preparation was treated with trypsin, as described in the text, prior to staining with Leishman. The chromosomes are differentiated into a pattern of light and dark bands, allowing unequivocal identification of each chromosome pair. [Photograph kindly provided by Dr M. Seabright, Wessex Regional Cytogenetics Unit, Salisbury]

(4) Stain in either 5% Giemsa (Gurr's R66 in phosphate buffer, pH 6.8) for 10 min or 2% Giemsa for 1.5 h. Rinse in distilled water and air dry.

IV.   Quinacrine (Q) Banding

Q-banding techniques were the first banding procedures to be described and were developed by Caspersson and his coworkers. Caspersson reasoned that the use of an alkylating agent, specifically one which attaches to the N7 atom of a guanine residue, could potentially be used to detect differences in base composition of DNA along a mitotic metaphase chromosome. If such an alkylating agent could be coupled to a fluorochrome, it was reasoned that the differences in base composition along a chromosome could be detected by examination of their fluorescence patterns. The alkylating agent which was used was quinacrine mustard dihydrochloride (Caspersson *et al.*, 1968). This chemical is fluorescent in the visible range, but in addition it demonstrates two kinds of interaction with DNA. The first is a non-specific intercalation into the double helix and the second is alkylation of the DNA bases, specifically the N7 atom of guanine. Another compound was also used, quinacrine dihydrochloride, which is also fluorescent and has the ability to intercalate non-specifically with DNA but not to alkylate the N7 atom of guanine. Initial experiments indicated that quinacrine mustard staining did indeed produce a pattern of brightly fluorescing bands along the length of the chromosome, but the same results were subsequently obtained using quinacrine dihydrochloride, which has no specificity for guanine residues (Caspersson *et al.*, 1968, 1969a, 1970).

In general, A–T-rich DNA enhances fluorescence, both on the chromosome and in solution (Weisblum and De Haseth, 1972), suggesting that Q-banding is specific for A–T-rich regions. However, in some cases A–T-rich DNA does not show bright fluorescence, as in the mouse, where the A–T-rich satellite DNA blocks fluoresce only dimly.

The very simplicity of the Q-banding techniques makes the mechanism of banding very difficult to analyse. Chromosome preparations are simply incubated in an aqueous solution of the dye, washed, and examined using fluorescence microscopy. Quinacrine dyes bind directly to the DNA, although by implication DNA–protein interactions must be involved in the production of the Q-bands because of their similarity to G-bands.

A. The 'Caspersson' method

This procedure was described by Caspersson *et al.* (1970).

(1) Chromosome preparations should be air dried from fixative or dehydrated in ethanol after squashing.

(2) Transfer the slides from absolute alcohol through an alcohol series (95%, 70%, 50%) into McIlvaine's buffer (disodium phosphate–citric acid buffer, pH 7.0, made by adding 35.3 ml of 0.1 M citric acid to 164.7 ml of 0.2 M disodium phosphate (Na_2HPO_4) in distilled water and adjusting the pH to 7.0).

(3) Stain in 50 μg/ml quinacrine mustard dihydrochloride in McIlvaine's buffer for 20 min at room temperature. Rinse quickly in three changes of buffer and mount in buffer. Seal the coverglass with nail varnish or rubber cement for examination. Examine using fluorescence microscopy using light of wavelength 480–500 nm. When exposed to ultraviolet light of this wavelength, the fluorescence can fade rapidly (a few minutes only). Therefore expose good, well stained chromosome sets to ultraviolet light for the mini-

mum possible time if a photographic record is required. References for microscopy and photography of Q-banded preparations are given at the end of the chapter.

B. Q-banding using quinacrine dihydrochloride

In this method (due to Uchida and Lin, 1974) the prestaining in step (2) above is omitted, the slide is immersed in the stain and washed as above.

(1) Prepare slides as above.

(2) Dip the slide in distilled water adjusted to pH 4.5 (with HCl) for 10 s.

(3) Stain in 0.5% quinacrine dihydrochloride in distilled water, pH 4.5, for 15 min.

(4) Rinse in three changes of water, pH 4.5, for a total time of 10 min.

(5) Mount in pH 4.5 water and seal the coverglass with nail varnish or rubber cement.

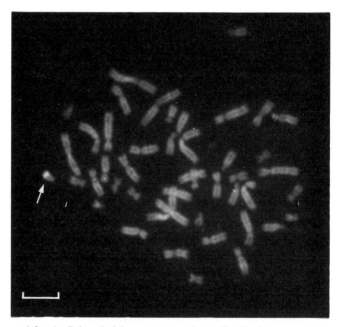

Figure 4.3   A Q-banded human metaphase. Each chromosome can be positively identified, and the Y chromosome is especially prominent due to the bright fluorescence of the long arm (arrow). Scale bar = 10 μm. [This photomicrograph was kindly provided by Lesley Barnett of The Department of Zoology, University of Leicester]

The staining solution will keep for at least a week if kept in the dark and refrigerated.

V. R-banding

Incubation of chromosome preparations in buffer at either a high temperature (Dutrillaux and Lejeune, 1971) or a suitable pH (Sehested, 1974) followed by Giemsa staining will produce a banding pattern that is opposite to that of G-banding and so is termed reverse or R-banding.

Very little has been done with regard to investigation of the mechanism of R-banding. However, by varying the temperature or pH of the treatment, G-banding will replace R-banding or *vice versa*, which suggests that the mechanisms of the two techniques are similar. This idea is supported by the fact that the R-banding pattern is the exact reverse of the G-banding pattern.

R-banding is of special interest where G-banding stains the ends of chromosomes lightly and R-banding darkly, as this makes visualization of chromosome ends much clearer. A typical example is in the human karyotype where both G- and R-banding are employed to determine whether there are deletions, or to pinpoint an exact break-point in the case of a translocation.

A.   R-banding by high temperature treatment

The following procedure is due to Dutrillaux and Lejeune (1971).

(1) Prepare chromosome preparations by any method; air drying is advisable.

(2) Incubate slides in 20 mM phosphate buffer, pH 6.5, for 10 min at 87 °C. Rinse with tap water.

(3) Stain in 2% Giemsa in phosphate buffer, pH 6.8, for 10–30 min.

B.   R-banding by low pH treatment

This low pH method was developed by Sehested (1974).

(1) Prepare chromosomes as above.

(2) Incubate slides in 1 M NaH_2PO_4 (unbuffered, this solution will have a pH of 4.0–4.5) at 88°C for 10 min. Rinse briefly in water (tap or distilled).

(3) Stain slides in 5% Giemsa in distilled water for 10 min.

C.   R-banding with acridine orange

A third method is that due to Verma and Lubs (1975).

(1) Prepare chromosomes by air drying, allow the slides to 'age' at room temperature for 7–10 days before processing.

(2) Place the slides in a Coplin jar containing fresh phosphate buffer (32 ml of 0.07 M $Na_2HPO_4.12H_2O$ plus 68 ml of 0.07 M KH_2PO_4, adjusted to pH 6.5 if necessary by adding more 0.07 M $Na_2HPO_4.12H_2O$) at 85°C and incubate for 20–25 min.

(3) Stain slides in 0.01% acridine orange in the above phosphate buffer for 4–6 min. Rinse in phosphate buffer and mount in same.

An excessive incubation in the hot phosphate buffer will result in the chromosomes appearing fuzzy and will produce an overall red fluorescence when viewed with ultraviolet illumination. Overstaining will also produce overall red fluorescence which can be corrected by additional rinsing. Understaining will produce green fluorescence which can be corrected by additional staining.

D.   Telomeric R-banding

Using this technique (Dutrillaux, 1973), the telomeres become strongly stained, but faint R-banding still occurs over the rest of the chromosomes.

(1) Prepare slides as normal, air dry, and allow to 'age' for a few days.

(2) Place slides in phosphate-buffered saline (PBS) for 20–60 min at 87°C but adjust the pH to 5.1. Rinse in PBS. Phosphate-buffered saline is made up as follows: 8 g NaCl, 0.4 g KCl, 4.23 g $Na_2HPO_4.7H_2O$, and 0.4 g KH_2PO_4, made up to 1 l and autoclaved. The pH can be adjusted to 5.1 with HCl or NaOH if necessary.

(3) Stain in 3% Giemsa in phosphate buffer, pH 6.8, at 87°C. Leave for 5–30 minutes. Rinse.

VI.   Silver staining

A staining technique has recently been developed which stains specifically the nucleolus organizer regions (NORs) (Goodpasture and Bloom, 1975; Howell et al., 1975). This technique, referred to as the Ag–AS stain, stains an acid protein component of NORs which are active, or which were transcriptionally active in the preceding interphase (D. A. Miller et al., 1976; O. J. Miller et al., 1976). The specificity of the staining technique is extremely high in mammalian material if the recommended procedure is rigorously followed. However, overstaining results in staining of the NORs plus the centromeres and ultimately the whole chromosome (Goodpasture and Bloom, 1975; Howell et al., 1977) and relatively slight modifications to the protocol result in staining of A–T-rich and satellite regions of human chromosomes (Howell and Denton, 1976), or of the centrioles (Howell et al., 1977). In some non-mammalian systems, staining may not be specific for

NORs. Such reports have been published for lampbrush chromosomes of *Triturus cristatus carnifex* (Varley and Morgan, 1978), for mitotic chromosomes of *Triturus* species (Ragghianti *et al.*, 1977), and for polytene chromosomes of *Rhynchosciara* (Stocker *et al.*, 1978) and *Chironomus* (J. M. Varley, unpublished observations).

The precise basis for the ammoniacal silver staining technique has still not been determined, but Goodpasture and Bloom (1975) put forward the following probable mode of action. During the initial incubation which is with a concentrated silver nitrate solution, the nucleolus and NORs are selectively impregnated. The following step is incubation in an ammoniacal silver solution and a formalin developer during which more metallic silver is deposited at the sites which have been selectively impregnated. The results with non-mammalian chromosomes indicate that the Ag-AS staining procedure may not only stain proteins associated with the NORs, but perhaps a family of closely related proteins, some of which may not be connected to the NORs at all, as suggested by Varley and Morgan (1978).

A.   Ag–AS staining

Three solutions are required for this staining technique, which is due to Goodpasture and Bloom (1975).

(a) *50% silver nitrate solution.* 10 g of AgNO$_3$ (Analar) dissolved in distilled water to a total volume of 20 ml. Filter the solution and store in a light-tight bottle at room temperature. It will keep indefinitely.

(b) *Ammoniacal silver solution (AS solution).* Dissolve 8 g of AgNO$_3$ in 10 ml of distilled water, then add 15 ml of concentrated ammonia solution (35% w/w NH$_3$, pH 12–13) slowly with stirring. Filter the solution and store in a light-tight bottle. This solution should always be made up at least 2 days before use, but not kept for longer than 2 weeks.

(c) *3% formalin.* A stock solution of 20% formalin should be made up and kept at 4°C. From this stock, dilute an aliquot to 3% final concentration using distilled water. Bring the pH to 7.0 with sodium acetate crystals, then adjust to pH 5.5 with formic acid. The pH of the 3% formalin determines the speed of differential silver staining: the lower the pH the faster the staining. Formalin is 40% formaldehyde; 20% formalin is made up by dissolving 8 g of paraformaldehyde in 100 ml of distilled water, raising the pH to 10 with 10 M NaOH, and heating until dissolved. Filter the solution through Whatman No. 1 paper, then neutralize with concentrated HCl.

The procedure itself may be described as follows:

(1) Chromosome preparations should be made by air drying, and 'aged'

for at least 2 days before use. Squashed preparations can be used, but generally with less success.

(2) Place three or four drops of 50% $AgNO_3$ on the preparation and cover it with a coverglass. Incubate the slide under a photoflood lamp (e.g. Philips 240 V, 500 W) so that the temperature of the slide reaches 60–70°C; this usually requires a distance between slide and lamp of 25–30 cm. Wait until the $AgNO_3$ starts to crystallize around the edge of the coverglass (10–15 min) and remove the lamp. Allow the slide to cool slightly, then rinse off the coverglass and $AgNO_3$ under running tap water. Air dry the slide.

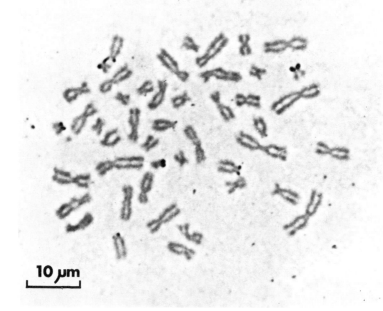

10 μm

Figure 4.4   An Ag–AS-stained human chromosome complement. In this preparation there was a faint yellow grainy background, the chromosomes were yellow, and the NORs were black. To enhance the contrast between the chromosomes and the background, this preparation was photographed using phase contrast. [Reproduced with permission from Varley, J. M., *Chromosoma (Berl.)*, **61**, 207–214, 1977]

(3) Mix two drops of AS solution with two drops of 3% formalin on a coverglass. If any silver deposit forms, discard and repeat. Mix quickly and lower the pretreated slide, preparation down, on to the coverglass. Immediately monitor the staining under the microscope, initially using phase contrast optics but, as staining proceeds, switch to bright field. Differential staining is achieved when the chromosomes turn pale yellow and the nucleoli and NORs are black (Figure 4.4), this usually takes 30–60 s. If staining is pro-

ceeding too fast, either raise the pH of the formalin slightly or use the AS solution and 3% formalin at 4°C instead of at room temperature. If preparations are understained, treatment with formalin and the AS solution could be repeated, but if this is done staining will progress very rapidly. If the NORs and nucleoli are black but the chromosomes themselves are faint, the preparation can be counterstained in Giemsa (2% Gurr's R66 in phosphate buffer, pH 6.8) or can be examined using phase contrast microscopy, in which case a halo will be seen around the silver deposits.

B.   A simplified method

This simplified method (Goodpasture and Bloom, 1975; Varley, 1977) is not suitable for all preparations. For example, lampbrush chromosome preparations of *Triturus cristatus carnifex* would not stain differentially using this procedure, but stained extremely well using the full method previously described (Varley and Morgan, 1978).

The protocol consists of incubating the preparation in silver nitrate alone.

(1) Chromosome preparations should be air dried and kept for at least 2 days before use.

(2) Place three or four drops of 50% $AgNO_3$ over the preparation and carefully lower a coverglass on to it. Prepare a moist chamber as described in Section II.B, step (6), but place distilled water on the filter paper. Place the slides in the moist chamber in an incubator at a temperature of 37–65°C.

(3) Incubate for 4–18 h, depending on the temperature. Lower temperatures require longer periods of incubation (up to a maximum of 18 h) to stain the NORs. Staining can be monitored periodically using phase contrast optics but great care must be taken not to touch the $AgNO_3$ with an objective or get $AgNO_3$ on any other part of the microscope. Silver staining using this method frequently requires counterstaining with Giemsa as described previously.

VII.   Demonstration of sister chromatid exchange

The first description of sister chromatid exchange was made by Taylor (1958), from studies in which plant cells were labelled with tritiated thymidine for one round of replication and then allowed to replicate once more without labelled thymidine. Autoradiographs of such preparations showed chromosomes in which only one of the two sister chromatids was labelled at any point, but occasionally the label switched from one chromatid to the other, demonstrating the occurrence of an exchange.

Techniques have now been developed which allow the unequivocal demonstration of sister chromatid exchanges without the use of a radioisotope.

The basic principle behind these techniques is that cells are grown up for two rounds of replication in the presence of a base analogue, 5-bromodeoxyuridine (BrdU). After two rounds of replication, one of the sister chromatids is unifilarly substituted with BrdU (i.e. only one polynucleotide strand contains BrdU), whereas the other is bifilarly substituted. The unifilarly substituted chromatids stain more darkly with Giemsa than bifilarly substituted ones (Zakharov and Egolina, 1973). The fluorescent stain Hoechst 33258 also stains the two chromatids differentially (Latt, 1973). A fluorescence plus Giemsa technique (FPG) is now most frequently employed to demonstrate sister chromatid exchange after BrdU substitution (Perry and Wolff, 1974; Wolff and Perry, 1974). The reason for differential staining after BrdU substitution seems to be that non-histone proteins are more tightly bound to DNA containing BrdU than to unsubstituted DNA and so substituted DNA is less densely packed and therefore stains lightly with Giemsa and fluoresces more brightly when stained with Hoechst 33258. Hoechst fluorescence fades very quickly and so Giemsa-stained preparations are preferable.

A.   The fluorochrome plus Giemsa (FPG) method

This technique was first developed in papers by Perry and Wolff (1974) and Wolff and Perry (1974).

   (1) It is essential to know the doubling time of the cells under examination in order to ensure that the cells are grown in the presence of BrdU for two rounds of replication. Once the doubling time is ascertained, grow the cells until they are growing exponentially and then add BrdU to a final concentration of 10 μM. Continue growing the cells until they have gone through two rounds of replication, but grow them in darkness to reduce the risk of increasing the frequency of sister chromatid exchanges due to photolysis. For some cells the concentration of BrdU may need to be adjusted. A concentration of 2 μM BrdU is recommended for some amphibian cells (H. Horner., personal communication).
   (2) Make air-dried chromosome preparations from the cells after colchicine block, hypotonic treatment, and fixation (see Chapter 2).
   (3) Stain the preparation in 0.5 μg/ml Hoechst 33258 in distilled water for 12–15 min; rinse in distilled water.
   (4) Mount in distilled water and seal the coverglass with nail varnish or rubber cement. Leave the slides for 24 h exposed to daylight (or another weak ultraviolet light source), during which time a photochemical reaction occurs which causes the chromatids to fluoresce differentially. If required, the chromosomes can be examined at this stage using fluorescence microscopy.

(5) After 24 h, remove the coverglass and incubate the slide for 2 h in either water or 2× SSC (if the latter is used, weak G-banding may be observed) at 60°C. Rinse briefly.

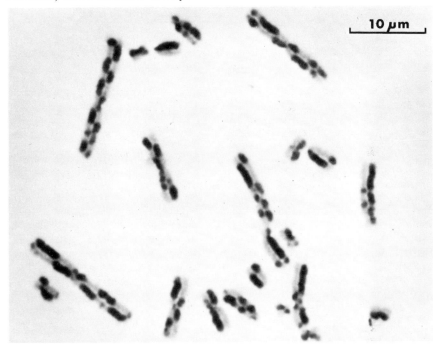

Figure 4.5 Mitotic chromosomes from a Chinese hamster tissue culture (CHO) cell showing multiple sister chromatid exchanges. The cells were grown in the presence of BrdU and the chromosomes were stained with Hoechst and Giemsa as described in the text. The multiple sister chromatid exchanges were induced by treatment of the cells with cyclophosphamide. (This photomicrograph was kindly provided by Dr Sheldon Wolff and Ms Judy Bodycote of the Laboratory of Radiobiology and Department of Anatomy, University of California at San Francisco, Calif.]

(6) Stain the slides in 3% Giemsa (Gurr's R66 in phosphate buffer, pH 6.8) for approximately 30 min. Rinse and mount if required.

On examination, the chromosomes should now exhibit a 'harlequin' appearance, demonstrating the occurrence of sister chromatid exchanges.

References

Alfi, O. S., Donnell, G. N., and Derencsenyi, A. (1973) C-banding of human chromosomes produced by DNase. *Lancet*, *ii*, 505.

92 *Working With Animal Chromosomes*

Arrighi, F. E., and Hsu, T. C. (1974). Staining constitutive heterochromatin and Giemsa crossbands of mammalian chromosomes. In *Human Chromosome Methodology* (J. J. Yunis, ed.), Academic Press, New York and London, Chap. 4.

Arrighi, F. E., Hsu, T. C., Pathak, S., and Sawada, H. (1974). Sex chromosomes of Chinese hamster – Constitutive heterochromatin deficient in repetitive DNA sequences. *Cytogenet. Cell Genet.*, **13**, 268–274.

Calderon, D., and Schnedl, W. (1973). A comparison between quinacrine fluorescence banding and ^{3}H thymidine incorporation patterns in human chromosomes. *Humangenetik*, **18**, 63–70.

Caspersson, T., Farber, S., Foley, G. E., Kudynowski, J., Modest, E. J., Simonsson, E., Wagh, U., and Zech, L. (1968). Chemical differentiation along metaphase chromosomes. *Exp. Cell Res.*, **49**, 219–222.

Caspersson, T., Zech, L., Modest, E. J., Foley, G. E., Wagh, U., and Simonsson, E. (1969a). Chemical differentiation with fluorescent alkylating agents in *Vicia faba* metaphase chromosomes. *Exp. Cell Res.*, **58**, 128–140.

Caspersson, T., Zech, L., Modest, E. J., Foley, G. E., Wagh, U., and Simonsson, E. (1969b). DNA-binding fluorochromes for the study of the organization of the metaphase nucleus. *Exp. Cell Res.*, **58**, 141–152.

Caspersson, T., Zech, L., and Johanson, C. (1970). Differential banding of alkylating fluorochromes in human chromosomes. *Exp. Cell Res.*, **60**, 315–319.

Comings, D. E., Avelino, E., Okada, T. A., and Wyandt, H. E. (1973). The mechanism of C and G banding of chromosomes. *Exp. Cell Res.*, **77**, 469–493.

Crossen, P. E., Pathak, D., and Arrighi, F. E. (1975). A high resolution study of the DNA replication patterns of Chinese hamster chromosomes using sister chromatid differential staining technique. *Chromosoma (Berl.)*, **54**, 339–347.

Dutrillaux, B. (1973). Nouveau système de marquage chromosomique: les bandes T, *Chromosoma (Berl.)*, **41**, 395–401.

Dutrillaux, B. (1975). Traitements discontinus par le BrdU et coloration par l'acridine orange: obtention de marquages R, Q et intermediaires. *Chromosoma (Berl.)*, **52**, 261–273.

Dutrillaux, B., and Lejeune, J. (1971). Sur une nouvelle technique d'analyse du caryotype humaine. *Compt. Rend. Acad. Sci.*, **272**, 2638.

Dutrillaux, B., Rethoré, M.-O., and Lejeune, J. (1975). Comparison of karyotype of orangutan (*Pongo pygmaeus*) to those of man, chimpanzee and gorilla. *Ann. Génét.*, **18**, 153–161.

Epplen, J. T., Siebers, J. W., and Vogel, W. (1975). DNA replication patterns of human chromosomes from fibroblasts and amniotic fluid cells revealed by a Giemsa staining technique. *Cytogenet. Cell Genet.*, **15**, 177–178.

Evans, H. J., Buckland, R. A., and Sumner, A. T. (1973). Chromosome homology and heterochromatin in goat, sheep and ox, studied by banding techniques. *Chromosoma (Berl.)*, **42**, 383–402.

Ganner, E., and Evans, H. J. (1971). The relationship between patterns of DNA replication and of quinacrine fluorescence in the human chromosome complement. *Chromosoma (Berl.)*, **35**, 326–341.

Geraedts, J. P. M., Pearson, P. L., Van Der Ploeg, M., and Vossepoel, A. M. (1975). Polymorphism for human chromosome I and chromosome Y: Feulgen and UV DNA measurements. *Exp. Cell Res.*, **95**, 9–14.

Goodpasture, C., and Bloom, S. E. (1975). Visualization of nucleolar organizer regions in mammalian chromosomes using silver staining. *Chromosoma (Berl.)*, **53**, 37–50.

Gosden, J. R., Mitchell, A. R., Buckland, R. A., Clayton, R. P., and Evans, H. J.

(1975). Location of four human satellite DNAs on human chromosomes. *Exp. Cell Res.*, **92**, 148–158.

Gosden, J. R., Mitchell, A. R., Seuanez, H. N., and Gosden, C. M. (1977). The distribution of sequences complementary to human satellite DNAs I, II and IV in the chromosomes of chimpanzee (*Pan troglodytes*), gorilla (*Gorilla gorilla*) and orangutan (*Pongo pygmaeus*). *Chromosoma (Berl.)*, **63**, 253–271.

Howell, W. M., and Denton, T. E. (1976). Negative silver staining in A–T and satellite DNA-rich regions of human chromosomes. *Chromosoma (Berl.)*, **57**, 165–169.

Howell, W. M., Denton, T. E., and Diamond, J. R. (1975). Differential staining of the satellite regions of human acrocentric chromosomes. *Experientia (Basel)*, **31**, 260–262.

Howell, W. M., Hsu, T. C., and Block, B. M. (1977). Visualization of centriole-bodies using silver staining. *Chromosoma (Berl.)*, **65**, 9–20.

Kim, M. A., Johanssman, R., and Grzeschik, K. H. (1975). Giemsa staining of the sites replicating DNA early in human lymphocyte chromosomes. *Cytogenet. Cell Genet.*, **15**, 363–371.

Latt, S. A. (1973). Microfluorometric detection of deoxyribonucleic acid replication in human metaphase chromosomes. *Proc. Natl. Acad. Sci.*, **70**, 3395–3399.

Latt, S. A. (1975). Fluorescence analysis of late DNA replication in human metaphase chromosomes. *Somatic Cell Genet.*, **1**, 293–322.

Miller, D. A., and Miller, O. J. (1972). Chromosome mapping in the mouse. *Science*, **178**, 949–955.

Miller, D. A., Dev, V. G., Tantravahi, R., and Miller, O. J. (1976). Suppression of human nucleolus organizer activity in mouse–human somatic hybrid cells. *Exp. Cell Res.*, **101**, 235–240.

Miller, O. J., Miller, D. A., Dev, V. G., Tantravahi, R., and Croce, C. M. (1976). Expression of human and suppression of mouse nucleolus organizer activity in mouse–human somatic cell hybrids. *Proc. Natl. Acad. Sci.*, **73**, 4531–4535.

Pardue, M. L., and Gall, J. G. (1970). Chromosomal localization of mouse satellite DNA. *Science*, **168**, 1356–1358.

Pathak, S., Hsu, T. C., and Arrighi, F. E. (1973). Chromosomes of *Peromyscus* (Rodentia, Cricetidae). 4. Role of heterochromatin in karyotypic evolution. *Cytogenet. Cell Genet.*, **12**, 315–326.

Perry, P., and Wolff, S. (1974). New Giemsa method for the differential staining of sister chromatids. *Nature (Lond.)*, **251**, 156–158.

Pierpont, M. E., and Yunis, J. J. (1977). Localization of chromosomal RNA in human G-banded metaphase chromosomes. *Exp. Cell Res.*, **106**, 303–308.

Ragghianti, M., Bucci-Innocenti, S., and Mancino, G. (1977). An ammoniacal silver staining technique for mitotic chromosomes of *Triturus* (Urodela: Salamandridae). *Experientia (Basel)*, **33**, 1319–1321.

Rowley, J. D. (1973). New consistent chromosomal abnormality in chronic myelogenous leukemia identified by quinacrine fluorescence and Giemsa stain. *Nature (Lond.)*, **243**, 290–293.

Seabright, M. (1971). A rapid banding technique for human chromosomes. *Lancet*, *ii*, 971–972.

Sehested, J. (1974). A simple method for R banding of human chromosomes, sharing a pH dependent connection between R and G bands. *Humangenetik*, **21**, 55–58.

Stock, A. D. (1976). Chromosome banding pattern relationships of hares, rabbits and pikas (order Lagomorpha). A phyletic relationship. *Cytogenet. Cell Genet.*, **17**, 78–88.

Stocker, A. J., Fresquez, C., and Lentzios, G. (1978). Banding studies on polytene chromosomes of *Rhynchosciara hollaenderi*. *Chromosoma (Berl.)*, **68**, 337–356.

Stockert, J. C., and Lisanti, J. A. (1972). Acridine orange differential fluorescence of fast and slow reassociating chromosomal DNA after *in situ* DNA denaturation and reassociation. *Chromosoma (Berl.)*, **37**, 117–130.

Sumner, A. T. (1972) A simple technique for demonstrating centromeric heterochromatin. *Exp. Cell Res.*, **75**, 304–306.

Sumner, A. T. (1974). Involvement of protein disulphides and sulphydryls in chromosome banding. *Exp. Cell Res.*, **83**, 438–442.

Sumner, A. T., and Evans, H. J. (1973). Mechanisms involved in the banding of chromosomes with quinacrine and Giemsa. II. The interaction of the dyes with the chromosomal components. *Exp. Cell Res.*, **81**, 223–236.

Sumner, A. T., Evans, H. J., and Buckland, R. A. (1971). A new technique for distinguishing between human chromosomes. *Nature New Biol.*, **232**, 31–32.

Sumner, A. T., Evans, H. J., and Buckland, R. A. (1973). Mechanisms involved in the banding of chromosomes with quinacrine and Giemsa. I. The effects of fixation in methanol–acetic acid. *Exp. Cell Res.*, **81**, 214–222.

Taylor, J. H. (1958). Sister chromatid exchanges in tritium labelled chromosomes. *Genetics*, **43**, 515–529.

Uchida, I. A., and Lin, C. C. (1974). Quinacrine fluorescent patterns. In *Human Chromosome Methodology* (J. J. Yunis, ed.), Academic Press, New York and London, Chap. 3.

Varley, J. M. (1977). Patterns of silver staining of human chromosomes. *Chromosoma (Berl.)*, **61**, 207–214.

Varley, J. M., and Morgan, G. T. (1978). Silver staining of the lampbrush chromosomes of *Triturus cristatus carnifex*. *Chromosoma (Berl.)*, **67**, 233–244.

Verma, R. S., and Lubs, H. A. (1975). A simple R-banding technique. *Amer. J. Human Genet.*, **27**, 110.

Weisblum, B., and De Haseth, P. (1972). Quinacrine – a chromosome stain specific for deoxyadenylate–deoxythymidylate-rich regions in DNA. *Proc. Natl. Acad. Sci.*, **69**, 629–632.

Weiss, M. C., and Green, H. (1967). Human–mouse hybrid cell lines containing partial complements of human chromosomes and functioning human genes. *Proc. Natl. Acad. Sci.*, **58**, 1104–1111.

Wolff, S., and Perry, P. (1974). Differential Giemsa staining of sister chromatids and the study of sister chromatid exchange without autoradiography. *Chromosoma (Berl.)*, **48**, 341–353.

Yunis, J. J., Kuo, M. T., and Saunders, G. F. (1977). Localization of sequences specifying messenger RNA to light-staining G-bands of human chromosomes. *Chromosoma (Berl.)*, **61**, 335–344.

Zakharov, A. F., and Egolina, N. A. (1972). Differential spiralization along mammalian mitotic chromosomes. I. BUdR-revealed differentiation in Chinese hamster chromosomes. *Chromosoma (Berl.)*, **38**, 341–365.

Additional references not cited in text

C-banding

McKenzie, W. H., and Lubs, H. A. (1973). An analysis of the technical variables in the production of C-bands. *Chromosoma (Berl.)*, **41**, 175–182.

Scheres, J. M. J. C. (1974). Production of C and T bands in human chromosomes after heat treatment at high pH and staining with 'Stains All'. *Humangenetik*, **23**, 311–314.

G-banding

Gormley, I. P., and Ross, A. (1972). Surface topography of human chromosomes examined at each stage during the ASG banding procedure. *Exp. Cell Res.*, **74**, 585–587.

Q-banding

Lin, C. C., Van De Sande, H., Smink, W. K., and Newton, D. R. (1975). Quinacrine fluorescence and Q-banding patterns of human chromosomes. I. Effects of varying factors. *Canad. J. Genet. Cytol.*, **17**, 81–92.

Rowley, J. D., Potter, D., and Mikuta, J. (1971). Reuse of chromosome preparations for fluorescent staining. *Stain. Technol.*, **46**, 97.

Schwarzacher, H. G. (1974). Fluorescence microscopy of chromosomes and interphase nuclei. In *Methods in Human Cytogenetics* (H. G. Schwarzacher, U. Wolf, and E. Passarge, eds), Springer Verlag, Berlin, Heidelburg, New York, Chap. V.

Sumner, A. T., Carothers, A. D., and Rutovitz, D. (1981). The distribution of quinacrine on chromosomes as determined by X-ray microanalysis. *Chromosoma (Berl.)*, **82**, 717–734.

Van Der Ploeg, M., and Ploem, J. S. (1973). Filter combinations and light sources for fluorescence microscopy of quinacrine mustard or quinacrine stained chromosomes. *Histochimie*, **33**, 61–70.

R-banding

Burkholder, G. D. (1981). The ultrastructure of R-banded chromosomes. *Chromosoma (Berl.)*, **83**, 473–480.

Sande, J. H. van de, Lin, C. C., and Jorgenson, K. F. (1977). Reverse banding on chromosomes produced by guanosine–cytosine specific DNA binding antibiotic, olivomycin. *Science*, **195**, 400–402.

Schweizer, D. (1977). R-banding produced by DNase I digestion of chromomycin-stained chromosomes. *Chromosoma (Berl.)*, **64**, 117–124.

Sister chromatid exchange

Kato, H. (1977). Spontaneous and induced sister chromatid exchange as revealed by the BUdR-labelling method. *Internat. Rev. Cytol.*, **49**, 55–93.

Wolff, S. (1977). Sister chromatid exchange. *Annu. Rev. Genet.*, **11**, 183–201.

General

Comings, D. E. (1978). Mechanisms of chromosome banding and implication for chromosome structure. *Annu. Rev. Genet.*, **12**, 25–46.

N-banding

Funaki, K., Matsui, S., and Sasaki, M. (1975). Location of nucleolar organizers in animal and plant chromosomes by means of an improved N-banding technique. *Chromosoma (Berl.)*, **49**, 357–370.

Matsui, S., and Sasaki, M. (1973). Differential staining of nucleolus organisers in mammalian chromosomes. *Nature (Lond.)*, **246**, 148–150.

5

Polytene chromosomes

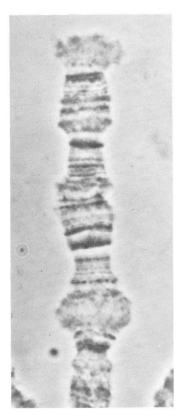

Drosophila melanogaster chromosome 3L showing puffing at the end and at division 63. Magnification 3500×. [This photomicrograph was kindly provided by Dr J. G. Gall of the Department of Biology, Yale University, New Haven, Conn.]

Polytene chromosomes were first described by Balbiani in 1881, but it was not until about 50 years later that more detailed studies revealed their true significance. Polytene chromosomes are much larger than mitotic or meiotic chromosomes. Polytene chromosomes from the salivary glands of *Drosophila melanogaster* are over 200 times the length of the mitotic chromosomes from the same species and the total length of the polytene chromosome complement can reach about 2 mm. The polytene chromosomes are not only extremely large, they also show linear differentiation into dark and light regions, termed bands and interbands respectively. The reproducibility of the banding patterns within one species and the characteristic patterns of different species were described by Painter (1933, 1934) and Heitz and Bauer (1933). Bridges (1935, 1936) went on to describe the altered banding patterns resulting from chromosomal rearrangements and he correlated these with phenotypic changes. Bridges suggested that the pattern of bands along the length of the polytene chromosomes represented individual genes and subsequently many workers correlated the order of polytene bands with linkage maps, their absence with gene deficiencies, and their duplication with other phenotypic effects.

Polytene chromosomes arise initially from normal somatic chromosomes which undergo several rounds of DNA replication without undergoing nuclear or cell division. The cells are therefore polyploid, but the chromatids remain laterally associated to produce the polytene type of organization. In addition to the association of the chromatids, there is frequently somatic pairing of the two homologous chromosomes which is at least as intimate as the pairing of homologues at meiotic prophase. Levels of polyteny may vary between different tissues in any one species, but the pattern of bands remains constant. In *Chironomus tentans*, the polytene chromosomes in the salivary gland may undergo up to 13 rounds of replication, and so contain 8192 (i.e. 2^{13}) laterally paired chromatids (Daneholt and Edström, 1967). The polytenes of the Malpighian tubules, however, only undergo nine rounds of replication (512 chromatids; data of Daneholt and Edström, 1967).

The polytene bands are regions which stain much more densely than adjacent interbands, although both contain DNA. Probably only about 5% of the total DNA in a *D. melanogaster* salivary gland nucleus is in the interbands and electron microscope studies have shown that the bands are composed of fibres densely packed into chromomeres. The bands vary considerably in width and it was originally suggested that the number of bands corresponds roughly to the number of genes as determined by mutation analysis (e.g. Judd *et al.*, 1972). However, it seems more likely that the number of genes exceeds the number of bands and that more than one structural gene may be present on one band. In *D. melanogaster* the average DNA content for a chromatid in a single band is 30 000 nucleotide pairs, with a range from 5000 to 200 000 nucleotide pairs (Rudkin, 1961, 1965;

Beermann, 1972), and there is evidence that a polytene band is a unit of genetic (e.g. Judd *et al.*, 1972; Lefevre, 1974) and of transcriptional activity (Beermann, 1972; Rudkin, 1972; McKenzie *et al.*, 1975). The band is also a unit of replication (Keyl, 1965).

At certain developmental stages, certain bands in the polytene set start to become diffuse, the tightly coiled chromomeres decondense, and the band becomes 'puffed'. The appearance of such puffed bands is constant within a tissue type at different developmental stages, but differs from tissue to tissue (Beermann, 1952; Ashburner, 1967, 1969a, b). The puffs are sites of intense RNA synthesis (Pelling, 1965) and the cytological manifestation of puffing represents the switching on of a gene (or genes) at a specific point in development, with regression of the puff indicating that the gene has been switched off. Stimuli found to induce puffing include heat shock and the hormones ecdysone and juvenile hormone (for a review see Zegarelli-Schmidt and Goodman, 1981).

Because of the endoreplicated state of polytene chromosomes in Dipteran larvae, it is possible to localize single gene sequences by *in situ* hybridization studies (Chapter 8, Section VI) or to localize proteins by immunofluorescence (this chapter, Section V.B). In *D. melanogaster* and other *Drosophila* species, the genetic map is well characterized and it has been possible to localize gene sequences to many polytene bands.

Polytene chromosomes are found in various tissues of the larvae of many Dipteran species. The most commonly used larval tissues are the salivary glands, but polytene chromosomes can also be found in Malpighian tubules, fat bodies, and gut epithelium; they are also present in ovarian nurse cells from the adults in some species (e.g. Beermann, 1952; French *et al.*, 1962). In general, salivary gland material is the easiest to handle and has the largest polytene chromosomes, but in some cases the polytenes from other tissues may be preferred (e.g. ovarian nurse cells in *Anopheles* spp., see Section VI).

In this chapter the preparation of polytene chromosomes from three genera, *Chironomus*, *Drosophila*, and *Anopheles*, will be discussed. *Chironomus* spp. provide probably the easiest material to work with and they have the advantage that they can be easily cultured or collected from the wild. *Drosophila* chromosomes are more difficult to work with and practice is required before consistently good preparations can be made. Salivary glands are the usual source of polytene nuclei in both the above genera, but in *Anopheles* the ovarian nurse cells are the preferred source. *Anopheles* has an obvious medical importance and polytenes can provide a good means of identifying and characterizing species that are otherwise morphologically indistinct.

I.   Principles of techniques for polytene chromosomes

No matter which tissue is used as a source of polytene chromosomes, the principles behind the techniques used for their isolation are the same. Polytene nuclei are always extremely large, as are the cells themselves, so the tissue in question is easy to manipulate and observe using at most a low power dissecting microscope. The tissue to be used must first be dissected out of the body; this can be accomplished either under a suitable balanced salt solution or in some cases using the haemolymph alone to moisten the tissues. Any extraneous tissue is dissected away from the organ, which is then fixed. A variety of fixatives can be used: the two that are most commonly used are ethanol–acetic acid (in the ratio 3 : 1) and 45% acetic acid. The fixed tissue is then transferred to a slide which has been coated with a thin film of gelatin (a 'subbed' slide, see Chapter 9, Section V) and is squashed after covering with a siliconized coverglass (see Chapter 3, Section II). The squashing procedure is generally carried out in 45% acetic acid; exceptions will be discussed later in the chapter. It is absolutely essential that the slide and coverglass are immaculately clean and dust free and that the tissue is clean of any attached material. The actual procedure for squashing the tissue requires a little practice. Basically the techniques involve initial rupture of the cell membrane followed by the nuclear membrane and subsequent squashing to flatten the chromosomes on the slide. The intensity of the squashing procedure and in some cases the whole technique may change according to the species used. Some closely related species may require different treatments in order to obtain good squash preparations of the polytene chromosomes. Covering a range of both minor and major variants of the squashing technique has been the aim of this chapter, but it must be emphasized that there is no standard technique for the preparation of polytene chromosomes from different species or tissues. Each individual laboratory employs a slightly different technique, so try a simple technique first and if that does not work adjust the conditions accordingly.

II.   Materials and equipment needed

(a) Dissecting microscope with bright incident light and a black stage plate
(b) Fine forceps (Watchmakers' No. 5 are ideal)
(c) Dissecting needles
(d) Pasteur pipettes
(e) Pencil with an eraser end
(f) Filter papers (about 11.0 cm diameter, size not critical)
(g) Watch glass
(h) Microscope with phase contrast and bright field optics
(i) Subbed slides, clean and dust free (see Chapter 9, Section V)

(j) Siliconized coverglasses, clean and dust free (see Chapter 3, Section II)

(k) Container for dry ice (insulated) and a flat metal block of about the size and weight of the base of a small retort stand or liquid nitrogen in a suitably insulated container

(l) Coplin jar containing 95% or 100% ethanol

(m) 45% acetic acid: glacial acetic acid and distilled water

(n) 3 : 1 ethanol–acetic acid, three parts 100% ethanol to one part glacial acetic acid, freshly made

(o) Insect Ringer: 7.5 g NaCl, 0.35 g KCl, 0.21 g $CaCl_2$ made up to 1 l then autoclaved in small aliquots

(p) Acetic-orcein: 0.5% orcein in 45% glacial acetic acid, heated to dissolve the orcein, cooled and filtered

(q) Lacto–aceto–orcein: 0.5% orcein in equal parts 85% lactic acid and glacial acetic acid. Heat mixture until orcein dissolves, allow to cool thoroughly then filter

(r) Lacto–acetic acid: one part lactic acid to one part acetic acid

(s) Giemsa: Gurr's R66 Giemsa, 5% of stock Giemsa solution in 50 mM phosphate buffer, pH 6.8

III. Polytene chromosomes from *Chironomus* salivary glands

It is probably easier to make polytene chromosome preparations from *Chironomus* salivary glands than from any other species. In addition *Chironomus* larvae are available all year round, in extremely large numbers from autumn to February. *Chironomus* can be collected from any pond, pool, or lake by sweeping the bottom with a large scoop net in regions under trees where leaves have fallen into the water, and sifting through the mud for the 'bloodworms'. After collection the larvae can be kept for a considerable time by maintaining them in some mud and pond water at 4°C.

In view of the ease with which chironomid larvae can be collected, maintained, and used to make good polytene chromosome preparations, they can be strongly recommended for use in a laboratory practical exercise, and a technique for making temporary preparations from chironomid larvae using an absolute minimum of equipment, thus reducing the cost and effort of setting up a practical class, will be described. *Chironomus* has also proved an extremely interesting research animal and its polytene chromosomes have been extensively studied. Although there is not as much known about the genetics of *Chironomus* as for *Drosophila*, it has often been used for studies of chromosome organization and gene expression. In *C. tentans* the fourth chromosome has three prominent puffs which have been termed Balbiani rings and the products of these puffs have been analysed (e.g. Edström *et al.*, 1980; Rylander and Edström, 1980; Rylander *et al.*, 1980). In a closely related species, *C. plumosus*, the small fourth chromosome has not usually undergone somatic pairing, so this species appears to have five chromosomes.

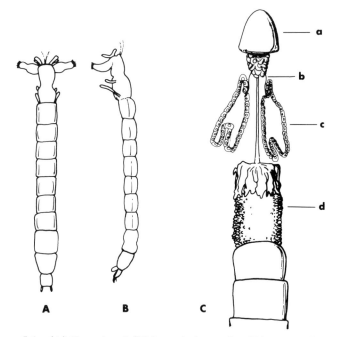

Figure 5.1   (A) Dorsal and (B) lateral views of a *Chironomus* larva. In both the head is at the bottom of the diagram. (C) Diagram of the appearance of the salivary glands after pulling off the head of a larva as described in the text: a, head; b, salivary duct; c, salivary gland; d, midgut

Polytene chromosomes are normally prepared from the salivary glands of chironomid larvae. The salivary glands comprise about 30 cells distributed around a central lumen (see Figure 5.1).

A.   Procedure

(1) To dissect out the salivary gland, remove a larva from the pond water and blot it dry gently on a piece of tissue. Place the larva on to a clean microscope slide and place this under a dissecting microscope with a black stage plate and a bright incident light. Grasp the larva with a pair of forceps about half way down the body. With a second pair of forceps grasp the head just behind the mouthparts and pull gently but firmly. Ideally the foregut will remain attached to the mouthparts and as the larva is decapitated the foregut, with the salivary glands attached to it laterally, will come cleanly

out of the body. If this does not happen, squeeze the contents out of the larva by laying the second pair of forceps just in front of where the larva is being held by the first pair and expel the body contents by running the flat edge of the forceps along the body towards the anterior end. The glands appear as translucent objects about 2–3 mm long (Figure 5.1).

(2) Clean away any debris from the glands and transfer them to a drop of 45% acetic acid on a clean, subbed microscope slide. This dissection can be carried out without the use of insect Ringer to keep the tissue moist, since there is adequate haemolymph to stop the preparation from drying out. The cleaned glands can be transferred to the 45% acetic acid using forceps, but since a lot of haemolymph is transferred between the forceps some workers prefer to transfer the glands using a dissecting needle. Very gently arrange the two glands so that they do not overlap and are lying fairly flat. Leave the glands in 45% acetic acid for at least 10 min, checking during this time that they do not dry out. A Petri dish lid can be placed over the slide to prevent dessication and to keep dust particles from landing on the slide.

(3) Thoroughly clean and dust a siliconized coverglass and lower it on to the preparation gently, until it touches the surface of the liquid. Take a piece of clean dry filter paper and fold it in half. Place the slide on one half and press the other half down on the slide to blot away any excess 45% acetic acid. Then gently but firmly press down on the coverglass, avoiding any horizontal movement. Look at the preparation under phase contrast using a low power objective. If most of the nuclei are still intact, then apply a little more pressure to the coverglass. Continue until the nuclei have burst and the polytene chromosomes are all lying in one plane. Monitor repeatedly under phase contrast and try not to oversquash the preparation. It is easy to apply more pressure if the preparation is insufficiently squashed, but there is no hope for an over-squashed preparation.

(4) As soon as the chromosomes are judged to be adequately spread (see, for example, Figure 5.2), make the preparations permanent as follows. Place the slide on a flat metal block standing on dry ice and leave for 10 min. Alternatively dip the slide in liquid nitrogen for about 15 s. At the end of this time, quickly flick the coverglass off the slide using a razor blade and plunge it immediately into a Coplin jar containing 95% or 100% ethanol. Leave for at least 30 min in the ethanol, then remove and air dry.

An alternative to the use of 45% acetic acid for fixing and squashing the salivary glands is to use a 0.5% solution of orcein in 45% acetic acid (with or without 0.5% carmine, which stains the nucleolus more strongly than does orcein) or in lacto–acetic acid. In this case the glands will be transferred directly into a drop of acetic–orcein, or lacto–aceto orcein, on a subbed slide and left for 20 min. After this time blot away any excess stain with a tissue, taking care not to touch the glands, and replace with a drop of 45% acetic

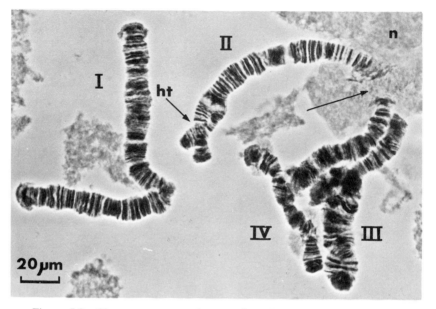

Figure 5.2 Chromosome complement of a salivary gland nucleus from a fourth instar larval of *Chironomus pallidivittatus*. The second of the four chromosomes (II) carries the nucleolus (n) and has a heterozygous region at one end (ht). The arrow indicates the gap produced in the chromosome by the nucleolus. The shortest chromosome (IV) has two large puffs (Balbiani rings). Several smaller puffs in other regions of the complement are also visible. The salivary glands were fixed for 3 min in ethanol–acetic acid (3 : 1) and stained for 1 h in acetic acid–orcein and acetic acid–carmine, 1 : 1. The cells were subsequently isolated in 50% acetic acid, squashed, frozen on dry ice, and mounted in Euparal. [This photomicrograph was kindly provided by Dr Ulrich Grossbach, Zoological Institute, University of Göttingen, Göttingen]

acid or lacto–acetic acid. Leave for about 2 min, blot away the excess liquid, replace with about 10 μl of 45% acetic acid or lacto–acetic acid and squash as for normal preparations. This procedure results in preparations which can be examined directly without treatment with dry ice and ethanol. Most analysis of polytene chromosome complements is done on preparations made in this way rather than those which are not stained during the squashing procedure.

To mount the preparations after staining, dip the slides in xylene for a few seconds, remove, drain off excess xylene, and place a drop of mounting medium (Permount, DPX, Euparal, etc.), and then blot away the excess mounting medium and xylene using a folded filter paper.

B. A quick method for *Chironomus*

A simplified technique can be employed to make temporary preparations suitable for examination in a laboratory practical class. Equipment required for this technique is minimal and is outlined below:

(a) Dissecting microscope
(b) Ordinary but very clean microscope slides and coverglasses
(c) Forceps or dissecting needles
(d) 45% acetic acid
(e) Acetic-orcein (0.5% orcein in 45% acetic acid)
(f) Filter papers
(g) Microscope with bright field optics
(h) Nail varnish or rubber cement

The basic method of preparation is identical to the one just outlined, but instead of using forceps to decapitate the larva, dissecting needles can be used. The dissecting needles should be laid across the body and behind the mouthparts, the larva should not be speared with them. If phase contrast optics are not available, the glands should be fixed in acetic-orcein as described previously, and the squashing monitored using bright field optics. Subbed slides and siliconized coverglasses are only necessary for permanent preparations, if anything squashing will be slightly easier if ordinary slides are substituted. After squashing, the preparation can be sealed by applying nail varnish or rubber cement around the edge of the coverglass to prevent drying out. If the polytenes are to be examined using high power objectives, be careful that the objectives do not touch wet nail varnish! Remember that although subbed slides and siliconized coverglasses are not necessary for temporary preparations, it is still essential that the slides and coverglasses are thoroughly clean and dust free.

C. Polytene chromosomes by Nomarski microscopy

Structures in the polytene nucleus such as the nucleolus and the Balbiani rings may not be readily apparent in preparations which have been fixed, stained, or dehydrated. One way of visualizing such features easily is by direct examination of the polytene cell by using phase contrast or Nomarski interference microscopy.

The technique for direct examination is identical to that for making permanent preparation as far as, and including, the actual squashing procedure. Once the tissue has been squashed sufficiently, seal the edge of the coverglass with nail varnish or wax and examine using phase contrast optics.

An alternative to the above technique is to prepare the salivary glands under liquid paraffin and to observe them using Nomarski interference optics. Blot a larva dry on a piece of filter paper, then place it in a small drop of liquid paraffin on an ordinary clean microscope slide. Decapitate the larva as described previously (Section III.A, step (1)), ensuring that the salivary glands are dissected out underneath the liquid paraffin. Remove as much debris as possible, then gently lower a coverglass on to the preparation. Do not squash the preparation until it has been examined using phase contrast or Nomarski optics. Very little squashing is required to give nuclei in which all the structures can be clearly seen. Excessive squashing results in a loss of resolution of the nuclear structures, and for Nomarski it is preferable to maintain the nucleus intact, albeit somewhat flattened (Figure 5.3).

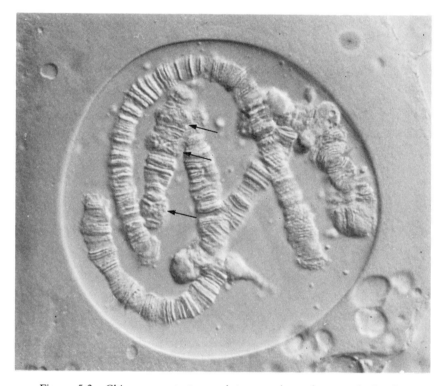

Figure 5.3 *Chironomus tentans* polytene nucleus photographed using Nomarski optics. The four chromosome pairs are clearly distinguishable and the Balbiani rings on chromosome IV are visible (arrows)

IV. Polytene chromosomes from *Drosophila*

A. Maintenance of stocks

It is easy to maintain a stock of live *Drosophila*, the flies are subcultured every 2 weeks and live on a medium that is simple to prepare.

Drosophila medium consists of yeast and glucose plus agar to give the correct consistency. In addition, a mould inhibitor is added. Most workers use a variation of one of two recipes, both of which are outlined below:

(a) *Alderson's medium.* 3% agar (Oxoid No. 3); 10% D-glucose; 10% yeast (dried); 0.06% Nipagin (Alderson, 1957).

(b) *Lewis's medium.* 0.45% agar; 8.3% corn meal; 5% dextrose; 2.5% sucrose; 1.5% dried yeast; 0.06% phosphoric acid; 0.4% proprionic acid (Lewis, 1960).

In both recipes all the ingredients except the mould inhibitors (nipagin/ phosphoric and proprionic acid) are autoclaved, the mould inhibitors are added to the cooled mixture, and aliquots are dispensed into suitable culture vessels. Wide necked bottles of about 500 ml capacity should be used, with a sponge stopper. Once the food has cooled sufficiently and is dry (i.e. no condensation around the sides of the bottles) about 20 flies can be placed in each bottle. Approximately 24 h after egg laying, the larvae will hatch out and can be seen as minute white larvae in and on the surface of the food. At this point the adult flies can be removed. *Drosophila* can be grown at temperatures between 18 and 25°C. *D. melanogaster* is frequently cultured at 25°C. Other species may prefer a lower temperature. Some workers do, however, prefer to grow *D. malanogaster* at a lower temperature, so trying a range of temperatures with any species being used is suggested, to obtain optimal growing conditions. Approximately 4 days after oviposition, there should be many third instar larvae present and these are the ones most commonly used for polytene chromosome studies. Third instar larvae will have moved off the surface of the food and crawled up the sides of the culture bottle. They are white and generally immobile, but will move if disturbed. Early prepupae, although still white or slightly brown, are completely immobile and can also be used.

B. Procedure

The salivary glands of *Drosophila* do not have the same structure as those of chironomids. They do not have a large central lumen and they contain two main types of cell. The large polytene cells are in the posterior part of the gland and much smaller cells lie towards the anterior end (Berendes and Ashburner, 1978). *Drosophila* salivary glands also have a fat body attached to them which must be removed before an attempt is made to squash the gland. They are also more difficult both to handle and to squash than the glands of the chironomids.

The polytene chromosomes of *Drosophila* larvae have some extremely interesting features. *Drosophila* polytene chromosomes are held together at a structure called the chromocentre, which is an aggregate of heterochromatin (Figure 5.4). During the polytenization steps, the heterochromatin at the chromocentre is underreplicated relative to the euchromatin, and has been termed 'α–heterochromatin' (Gall et al. 1971; Heitz, 1934a,b,; Rudkin, 1969). There are regions next to the chromocentre which are also condensed, but which are not underreplicated, and these have been termed 'β-heterochromatin' to distinguish them from both the 'α-heterochromatin' of the chromocentre and from the euchromatin. The term 'β-heterochromatin' is, however, something of a misnomer, as this material has a banded structure, is transcribed, and replicates with the euchromatin. It may therefore consist of euchromatin whose structure is affected by the proximity of the 'α-heterochromatin', rather than being heterochromatin itself.

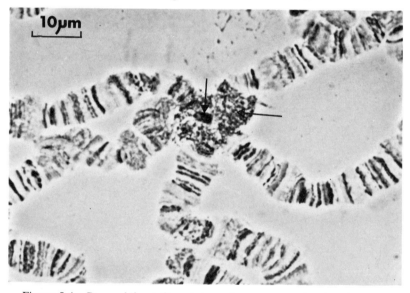

Figure 5.4 *Drosophila virilis* polytene chromosomes showing (solid arrow) the α-heterochromatin (open arrow), the β-heterochromatin, and the bases of the chromosome arms [Reproduced with permission from Gall, J. G., Cohen, E. H., and Polan, M. L., *Chromosoma (Berl.)*, 33, 319–344 (1971)]

Accordingly, the polytene chromosome complement in *Drosophila* species appears as a number of long banded chromosome arms radiating from the chromocentre (Figure 5.4). The convention used in nomenclature of the chromosome arms and bands follows that of Bridges (1935), and references to examples of maps of several *Drosophila* species are given at the end of this chapter.

Polytene chromosomes can also be isolated from salivary glands of larvae earlier than third instar, from prepupae, from pupae (e.g. Ashburner, 1967, 1969a, b), and from certain adult tissues. A protocol will be described in detail for the preparation of polytene chromosomes from the salivary glands of late third instar larvae of *D. melanogaster* which can be used for any other stage with little modification. The technique is adapted from that of Atherton and Gall (1972).

(1) Remove a late third instar larva from the side of the culture bottle and wash it in a little insect Ringer in a watchglass. Then place it in a drop of insect Ringer on an ordinary clean microscope slide under a low power dissecting microscope. Adjust the light source to give bright incident light shining on to a black stage plate.

(2) Hold the posterior end of the larva with one pair of forceps and with a second pair grasp the head immediately behind the black mouthparts. Pull the head gently but firmly and it should come cleanly away from the body with the paired salivary glands attached to it by the salivary ducts. If the glands do not come away with the head, gently squeeze the body from posterior to anterior to expel them. If pupae are used, the glands will not be attached to the mouthparts and so must be squeezed out of the body. The glands themselves should not be touched at this stage, but they can be manipulated by holding either the ducts or the fat bodies. Both larvae and white prepupal larvae have very fragile glands, and so subsequent stages must be carried out with great care.

(3) Gently tease the glands with the associated duct and fat bodies away from the rest of the viscera, then remove as much of the fat body as possible. The best preparations will be made from the glands with the least amount of fat body attached, but do not risk damaging the gland in order to remove all the fat body.

(4) Once the glands are as clean as possible they should be transferred to a watchglass containing freshly made 3 : 1 ethanol–acetic acid. They can be transferred by grasping the duct with a pair of forceps, or by picking them up on a dissecting needle. Leave the glands in the fixative for at least 2 min.

(5) Transfer the glands to a 10 μl drop of 45% acetic acid on a siliconized coverglass. They should be transferred with the minimum amount of fixative and this is most easily accomplished by picking them out of the fixative and watching them under the dissecting microscope until most of the fixative has evaporated. Alternatively, put the glands on to a clean, dry, siliconized coverglass straight from the fixative and wait until most of the fixative has evaporated. When it has, immediately add 10 μl of 45% acetic acid. In either

case, if the glands are not completely submerged in the 45% acetic acid push them under the surface using the forceps or a dissecting needle. Examine under the dissecting microscope and adjust the two lobes of the gland so that they are not overlapping at all.

(6) Gently lower a clean, dry, subbed slide on to the coverglass until it just touches the drop of acetic acid, avoiding air bubbles, and then invert the slide so that the coverglass is on top.

(7) The next part of the procedure is the most difficult, and may require some practice. It is essential to monitor the squashing of the preparation under phase contrast optics constantly during the procedure in order to attain the optimal conditions for spreading the chromosomes. Gently disrupt the tissue by moving the coverglass horizontally using either the tip of a dissecting needle or the eraser end of a pencil. Do not apply downwards pressure at this stage. The purpose of the horizontal movement of the coverglass is to free the nuclei from the salivary gland cells and allow the cytoplasm to flow freely. If it is difficult to move the coverglass or if, on examination by phase contrast microscopy, there is little or no cytoplasmic flowing, add a little more 45% acetic acid to the edge of the coverglass and repeat the horizontal movement. Continue moving the coverglass until the chromosomes are being freed from the disrupted nuclear envelope and all the cytoplasm has dispersed well.

(8) Place the slide, coverglass up, between a folded filter paper disc and gently blot away any excess 45% acetic acid. Then tap the coverglass sharply with the eraser end of a pencil some 5 to 10 times. Check under phase contrast; the chromosomes should now be spreading well. Do not allow any sideways motion while tapping the coverglass as at best this will cause distortion of the chromosomes, and at worst will cause chromosome breakage. If necessary tap the coverglass a few more times until the desired degree of spreading is reached. Do not attempt to have every chromosome set spread well; this rarely if ever occurs, and it is better to stop when a few sets of chromosomes are well spread rather than spoil these by trying to spread all the nuclei to the same quality. An ideal preparation will not only have well spread chromosome arms, but the chromosomes themselves will be flat. Check this by examining under a higher magnification (40× objective). If the chromosomes are in focus in a single plane, then the correct amount of spreading has been achieved. If they are not, then tap a little harder with a firm vertical motion; check and repeat until the chromosomes are flat. It should now be obvious that at all stages in spreading the polytene chromosomes it is essential to monitor the degree of spreading and flattening using phase contrast optics, and also that if the slide or coverglass were dirty, then there is no possibility of producing a well spread, flat, chromosome preparation.

(9) To increase the degree of flattening of the squash preparation, the

slide can now be placed on a 45°C heating block for 2–5 min. This step is optional, but is commonly used if preparations are to be used subsequently for *in situ* nucleic acid hybridization, or other fine structure studies. Again, during this heating step monitor regularly under phase contrast. The purpose of this step is to evaporate residual solvent and to increase the flattening of the chromosomes, but if it is exceeded then the chromosomes will rapidly degenerate into a non-refractile mass. Check particularly the state of the chromocentre; once it starts to lose its morphology remove the slide immediately.

(10) Whether or not the heating step has been included, now place the slides on a flat metal block on some dry ice and leave for 10–15 min, or dip into liquid nitrogen for about 15 s.

(11) Remove the slides from dry ice and immediately flick the coverglass off by prising a razor blade under one corner. Make sure that this is done in one swift movement and that the coverglass does not fall back on to the preparation. Plunge the slide directly into a Coplin jar containing 95% or 100% ethanol, leave for at least 30 min, remove, and air dry.

(12) Slides can be stored in air tight boxes at 4°C (preferably with a small bottle of a standard dessicant in the box), until required. Slides can be kept like this for at least 8 weeks.

The above method may not be suitable for all species of *Drosophila* and if difficulty is experienced in breaking open the nuclei and/or dispersing the chromosomes well, the problem may be solved by incorporating an acid fixation step into the protocol. One method which is widely used is that of Yoon *et al.* (1973). The dissection of the salivary glands is as described above, but instead of fixing in 3 : 1 ethanol–acetic acid, the glands are transferred immediately into 45% acetic acid for 5–10 s and then into 1 M HCl for a further 30 s, transferring the minimum amount of acetic acid. The glands can then either be squashed as described above (steps (5)–(12)) or they can be stained with orcein.

Polytene chromosome spreads can be stained after air drying using 2% Giemsa in Sorensen's buffer, pH 6.8, but more commonly they are stained using either acetic-orcein or lacto-aceto-orcein during their preparation.

The same procedure is followed for staining using either acetic-orcein or lacto-aceto-orcein. After the glands have been dissected out and cleaned up, transfer them to a clean, dry, subbed slide with a drop of the stain on it. Leave for about 20 min, ensuring that the preparation does not dry out. Blot away the excess stain, and replace with 45% acetic acid (if acetic–orcein was used) or lacto–acetic acid (if lacto-aceto-orcein was used). Blot away the excess after 2 min and repeat twice more. Finally cover the glands with 10 μl fixative and a siliconized coverglass and squash as described (steps (7)–

(9)). Seal the preparation with nail varnish or rubber cement and examine using bright field optics and a 16× and 40× objective.

V. Polytene chromosomes for looking at chromosomal protein

The fixatives described for use with salivary gland preparations are perfectly adequate if the polytenes are to be examined directly or if they are to be used for *in situ* nucleic acid hybridization studies. However, both acetic acid–ethanol and 45% acetic acid extract large amounts of protein from the chromosomes within a very short time. In particular the total histone content of the polytenes is reduced dramatically, especially that of histone HI (Dick and Johns, 1968). For some subsequent examinations of the polytene chromosomes, notably immunofluorescent studies of the distribution of histone and non-histone chromosomal proteins, it is essential to reduce the level of depletion of proteins. The problem can be reduced by the use of 45% acetic acid, 10 mM $MgCl_2$, 3.3% formaldehyde as a fixative (Silver and Elgin, 1976). However, this procedure still results in the extraction of most histones and about 10% of the non-histone chromosomal proteins, a result of the faster penetration of acetic acid into the cells (Chalkley and Hunter, 1975). Several groups have developed methods for preparing polytene chromosomes where only a minimal amount of protein is removed (Plagens *et al.*, 1976; Silver and Elgin, 1976, 1978; Silver *et al.*, 1978). The principle of all of these techniques is the same. Glands are dissected out in an isotonic buffer and are then fixed in a medium containing formaldehyde. At some stage the glands are incubated in medium containing formaldehyde plus a non-ionic detergent to solubilize cytoplasmic membrane structures. A comprehensive discussion of the principles of such a technique is given in Silver *et al.* (1978) and the following protocol has been taken from this source with additional modifications from Dr S. C. R. Elgin.

The technique was developed specifically to allow identification of chromosomal proteins and their distribution *in situ* using fluorescent labelled antibodies to those proteins as probes. The basic technique for the preparation of polytene chromosomes in this way can be used for any other subsequent examination of chromosomal proteins. The procedure given below refers to *Drosophila* species, but is also suitable for other Diptera.

A. Procedure

(1) Third instar larvae (raised at 25°C for 4–5 days) are removed from the side of the culture bottle and are washed in an isotonically balanced medium (Cohen and Gotchel, 1971; medium G): 25 mM disodium glycerophosphate; 10 mM KH_2PO_4; 30 mM KCl; 10 mM $MgCl_2$; 3 mM $CaCl_2$; 162 mM sucrose, pH 6.8; 0.5% Nonidet P40 (BRL).

(2) The salivary glands are dissected out in medium G plus 0.5% Nonidet

P40. Take care not to damage the glands in any way and do not attempt to remove the associated fat bodies. If the gland becomes cloudy at this stage it has been damaged and should be discarded. Incubate the gland in this solution for 10 min, during which time the cytoplasmic membranes will be solubilized, but the nuclei will remain intact. There will be no extraction of chromosomal proteins by the detergent.

(3) Transfer the glands to a fixative solution containing Nonidet P40 and formaldehyde: 100 mM NaCl; 10 mM $MgCl_2$; 2% Nonidet P40; 2% formaldehyde; 10 mM sodium phosphate buffer, pH 7.3. Incubate at room temperature for a maximum of 30 min. A fixation time of 30 min will obscure some antigenic determinants in the denser chromomeres.

(4) Transfer the glands to 45% acetic acid containing 10 mM $MgCl_2$ and dissect away the fat bodies. After 2 min or more, transfer the glands to a 10 μl drop of 45% acetic acid, 10 mM $MgCl_2$ on a siliconized coverglass and squash as previously described for normal preparations (see Section IV for steps (5)–(8) and (10)). Leave the gland in 45% acetic acid and 10 mM $MgCl_2$ for at least 10 min before squashing.

(5) After squashing, the slides can be treated in one of two ways. If the slides are to be stored, they should immediately be placed in storage medium (67% glycerol and 33% TBS (TBS is Tris-buffered saline: 0.01 M Tris, pH 7.15 (at 25°C) and 0.15 M NaCl)) after the coverglass has been flicked off. Slides can be stored in this medium at −20°C for up to 10 days. If the slides are to be used immediately, they should be plunged into TBS as soon as the coverglass has been flicked off. The slides can be kept in TBS at 4°C until required.

In our experience it is more difficult to obtain polytene squash preparations of good quality using the above method than with the normal techniques and so more practice may be required to get consistently good results. For some subsequent treatments, such as staining of chromosomal proteins, it may be adequate to plunge the slides in ethanol after the dry ice or nitrogen treatment, followed by air drying prior to staining.

B. Indirect immunofluorescent staining

Polytene chromosomes prepared in the way just described are suitable for many experimental manipulations. The chromosomal proteins are left virtually unaltered on the polytenes, with less than 1% of histones or non-histone chromosomal proteins lost. The chromosomes can be used for *in situ* hybridization, for staining of chromosomal constituents (including proteins), and for immunofluorescent studies. The last-mentioned technique is an extremely sensitive and powerful way to localize chromosomal constituents and has been used to localize many types of molecule, including RNA polymerases, histones, and conformational variants of histones and DNA. The principle of the technique is that an antibody is raised to the constituent in question using standard techniques. This antibody is then incubated with the polytene chromosome preparation, where it will bind specifically to the complementary antigen, i.e. the original constituent used to raise the antibody. There now follows an incubation with a second antibody, raised in a second species, against the IgG fraction of the first animal. This second antibody is chemically coupled to a fluorescent dye, fluorescein isothiocyanate (FITC), and will bind specifically to the first antibody. When chromosome preparations are treated in this way and are viewed using ultraviolet light, the position of the original chromosome constituent will be revealed by the fluorescence of the FITC-coupled antibody.

Full details of the preparation of antigens and antisera for use with polytene chromosomes are given in Silver *et al.* (1978), and the technique has also recently been modified for use with monoclonal antibodies. No details of production of the antigens and antisera will be given here, but a procedure for indirect immunofluorescent staining will be given.

(1) Remove the slides from the storage medium or TBS and wash three times for 7 min each in TBS at 22–24°C. Do not allow the squash to dry out at any stage – this is critical.

(2) Remove the slide from the last TBS wash and quickly blot dry the regions to either side of the actual squash using filter paper or a paper towel. Place each slide on a support in a moist chamber (see Chapter 6, Section III.B) containing water-soaked paper.

(3) Place 100–250 μl of an appropriate dilution of an antiserum quickly on each squash. Suitable dilutions of antiserum range from 1 : 2 to 1 : 4 and are in TBS containing 1–2 mg/ml of bovine γ-globulin (Sigma) to reduce non-specific staining. The dilution of the antiserum will depend on the titre and must be determined for each sample (see Silver *et al.*, 1978). Incubate for 15–30 min at 22–24°C.

(4) Rinse three times for 5 min each in TBS at 22–24°C.

(5) Perform the second antibody incubation as described above using a 1 : 20 to 1 : 50 dilution of the FITC-conjugated IgG fraction of goat anti (rabbit γ-globulin) serum (Miles Laboratories) in TBS. This operation should be carried out in a darkroom using a red light.

(6) Wash the slides three times for 5 min each in TBS at 4°C. It is not necessary to carry out these washes in the dark, but it is essential to shield the preparations from fluorescent lights.

(7) Place a drop of 9 : 1 glycerine–1 м Tris-HCl (pH 8.1) on to each preparation (while still wet), drop a clean coverglass on to it, and place the slide in the fold of a piece of folded filter paper. Apply intensive thumb pressure on the coverglass and view the slide using ultraviolet illumination.

Slides may be stored in the dark at 4°C for several months before viewing. For examination of the preparations, a narrow band exitation filter (480–490 nm) should be used, preferably in conjunction with an epi-illumination optical system. Suitable systems are marketed by several companies, including Zeiss and Leitz. A 40× Neofluar objective is suitable for examination of polytenes treated in the above way. Typical photographic exposure times using standard films (e.g. Ilford FP4, Kodak Tri-X) are in the order of 1.5– 2 min.

VI.   Polytene chromosomes from the nurse cells of mosquitoes

The majority of studies on mosquitoes have been carried out on *Anopheles* for two main reasons. First, some anophelene mosquitoes are vectors of human malaria, and secondly, good quality polytene chromosome preparations can only be obtained from *Anopheles* species. Many *Aedes* and *Culex* species are also important disease vectors, but cytogenetic studies are complicated by the difficulty in preparing polytenes from their tissue. Polytene chromosomes are present in several larval tissues, e.g. salivary glands, Malpighian tubules, and mid- and hindgut. Larval salivary glands have also been extensively studied in a wide variety of species. However, in some anophelene mosquitoes, polytene chromosomes develop transiently in the adult ovarian nurse cells (Coluzzi, 1968), and they may have better polytenes than larval tissues. An added advantage of the nurse cell source is that females caught in the wild can be dissected immediately and squashes made with the minimum of equipment. There is a considerable amount of sibling speciation in *Anopheles* and so there is little morphological evidence for the taxonomic separation of, for example, vectors and non-vectors of diseases. There is no easy morphological basis for identification of many species and in such cases the cytogenetic study of the polytene chromosome complement is an invaluable aid. Prior to the use of adult ovarian nurse cells, females captured in the wild were cultured, and examination of the salivary glands of their larval progeny undertaken. Details of the technique for preparing polytene squashes from larval salivary glands are given in French *et al.* (1962), but this technique does not differ significantly from those already described for chironomids and *Drosophila*.

The usefulness of the study of ovarian polytene chromosomes in studies of anophelene biology is evident from the studies of the *Anopheles gambiae* complex. Identification of the members of this species complex was accomplished by demonstration of hybrid sterility barriers between species. Studies of the nurse cell polytene chromosomes confirmed these data, but in addition a new member of the species group was identified (reviewed by White, 1981a, b). Identification of the members of this species complex is now routinely carried out using cytogenetic techniques, as the pattern of polytene bands on the X chromosome differs between species. The evolutionary relationships between the *Anopheles gambiae* complex species can be deduced from the differences in the polytene banding patterns which seem to have arisen predominantly as a result of a limited number of inversions (Coluzzi *et al.*, 1979). Inversion polymorphisms within the individual species are also important and these show a non-random distribution both geographically and in relation to local environmental effects such as humidity.

Perhaps one of the main areas of research into anophelene cytogenetics at this time relates to the potential genetic control of disease vectors by chromosome manipulation including the formation of translocation stocks (Curtis, 1980). Anophelene mosquitoes do, however, have many other potentially interesting features, such as the occurrence and importance of inversion polymorphisms, especially with regard to the fitness of the inverted or standard homozygotes, and cytogenetic analysis of taxonomic relationships.

A. Maintenance of stocks

Studies of mosquitoes can be carried out both on female flies captured in the wild and studied immediately and on animals that have been mass cultured. A brief outline of a technique for mass culturing of *Anopheles stephensi* and for preparation of polytene chromosomes from adult ovarian nurse cells will be given. We are grateful to Dr C. P. F. Redfern for providing us with technical details for these protocols.

Adult *Anopheles stephensi stephensi* can be mass cultured in cubic netting cages of 15–30 cm^3. They should be maintained at a temperature of 25°C with a 12 h photoperiod and an 80% relative humidity. If humidifiers are unavailable, small scale cultures can be maintained by placing cages in a large Perspex box containing several trays of saturated sodium chloride. The adults should be kept at a density of 50–300 adults per cage. Adults should be fed on a 10% w/v sucrose solution via absorbent paper wicks, but *A. stephensi* is an anautogenous mosquito and females will only develop and lay eggs after a blood meal. As the presence of sucrose within the ventral diverticulum may limit the size of the blood meal and, consequently, the number of developing eggs, it is preferable to starve females for 12–24 h before the blood meal. Anaesthetized guinea-pigs or mice are used for blood-feeding. A moderate-sized guinea-pig (about 500 g) can be anaesthetized with an intraperitoneal injection of 0.2–0.3 ml sodium phenobarbitol (60 mg/ml 'Sagatal'). Shaving the ventral surface of the guinea-pig will aid successful feeding. If guinea-pigs are used regularly, the dose of Sagatal will have to be increased up to a maximum of 1 ml, at which point the guinea-pig should no longer be used. Ideally mosquitoes should be blood-fed at least once a week. Investigators using guinea-pigs for this purpose in the UK require a Home Office Licence for experiments on live animals. Eggs are laid on the surface of water 2–3 days after feeding and a small dish of water lined with filter paper should be provided.

After the eggs have hatched, transfer the aquatic larvae to large plastic bowls of water. The larvae should be at an uncrowded density of 0.5 larvae per square centimetre of water surface and maintained at 25°C. 'Conditioned' tap water that has been allowed to stand at room temperature for a few days to encourage growth of microorganisms can be used to incubate the larvae. First to third instar larvae should be fed on a 1 : 1 mixture of neutralized liver digest (Oxoid) and ground Farex (baby cereal available commercially in the UK), by sprinkling the mixture over the surface of the water. Fourth instar larvae can be fed on ground Farex alone. (A suitable substitute for Farex in the USA is Gerber's mixed cereal pablum.) The best results are obtained by feeding twice a day, using small quantities at each feed to reduce the amount of food left over. Take great care to keep the surface of the water free from scum, which will form if too much food is given. Larvae are surface feeders and if a scum does form, they will be unable to feed or to obtain air.

Pupae can be collected either individually, using a wide mouthed pipette, or in bulk, by using a separating funnel. Collect all larvae and pupae in a sieve and transfer to a separating funnel containing cold tap water. The pupae will float and congregate at the surface if not disturbed, while the larvae sink and can be run out of the funnel. The pupae should then be placed in a small dish and allowed to eclose into a netting cage.

B. Procedure

The polytrophic ovarian follicle of *Anopheles stephensi* consists of seven nurse cells and one oocyte, surrounded by a layer of epithelial cells. After eclosion, the primary follicle in each ovariole develops to a resting stage at which the oocyte forms half or less of the follicle volume. At this stage the nurse cell polytene chromosomes are small, tightly coiled, and of poor cytological quality. Synchronous maturation of all the primary follicles is initiated by blood-feeding, and good polytene chromosome preparations can be made from 12 to 36 h after a blood meal. If females have been caught in the wild, those caught in mid-morning will be most likely to yield the best polytenes, otherwise they can be kept at 25°C for 3–12 h as necessary.

Adult post-blood-meal females should be stunned by the use of pyrethrum spray. Chloroform and ether should not be used as excessive exposure may result in penetration of the solvent into the ovaries, with the result that the nurse cell nuclei will be a dense, non-spreading mass.

Dissect the ovaries in saline (e.g. Becker's saline; Becker, 1959) by pulling off the two terminal abdominal segments. It may be advisable to pull off the legs and wings before starting the dissection. If the ovaries are not attached to these two segments, gently squeeze the body using a dissecting needle. Fix the ovaries in 3 : 1 ethanol–acetic acid for about 1 min. Transfer the ovaries to a drop of 45% acetic acid on a siliconized coverglass, allow the tissue to swell for a few seconds, and then dissect the individual nurse cell bundles with tungsten needles. Maturing oocytes contain a lot of protein and fatty material which results in a lower quality preparation than if the nurse cell bundles are separated and used alone. Accumulate several nurse cell bundles and transfer these in a drop of 45% acetic acid on to a clean, siliconized coverglass (the use of a siliconized Pasteur pipette is recommended at this stage). Gently lower a clean, subbed slide on to the coverglass and gently squash the preparation using thumb pressure alone. Check using phase contrast optics and, if the preparation is satisfactory, make the preparation permanent using the dry ice–ethanol method. The preparations can alternatively be made using acetic-orcein to stain in place of 45% acetic acid, as described previously (Section III.B).

Preparations can also be sealed with nail varnish after squashing if dry ice is not available. This is especially useful is squash preparations are being made in the field, where facilities are minimal.

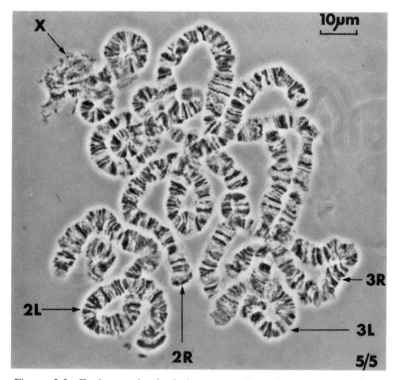

Figure 5.5   Feulgen-stained whole nurse cell nucleus from *Anopheles stephensi stephensi*, showing the three pairs of chromosomes, X, 2, and 3. [Reproduced by permission of Dr C. Redfern]

Anophelene mosquitoes have three pairs of chromosomes, males being the heterogametic sex.

Recent reviews of aspects of mosquito cytogenetics plus references to maps of many species are given in White (1981a, b).

References

Alderson, T. (1957). Culture conditions and mutagenesis in *Drosophila melanogaster*. *Nature (Lond.)*, **179**, 974–975.

Ashburner, M. (1967) Autosomal puffing patterns in a laboratory stock of *D. melanogaster*. *Chromosoma (Berl.)*, **21**, 398–428.

Ashburner, M. (1969a). The X chromosome puffing pattern of *D. melanogaster* and *D. simulans*. *Chromosoma (Berl.)*, **27**, 47–63.

Ashburner, M. (1969b). Patterns of puffing activity in the salivary gland chromosomes of *Drosophila*. IV. Variability of puffing patterns. *Chromosoma (Berl.)* **27**, 156–177.

Atherton, D., and Gall, J. G. (1972) Salivary gland squashes for *in situ* nucleic acid hybridization studies. *Drosophila Inf. Serv.*, **49**, 131–133.

Balbiani, E. G. (1881). Sur la structure du noyau des cellules salivaires chez les larves de *Chironomus*. *Zool. Anz.*, **4**, 637–641, 662–666.

Becker, H. J. (1959). Die puffs der Speicheldrüsenchromosomen von *Drosophila melanogaster*. I. Beobachtungen zum Verhalten des Puffmusters im Normalstamm und bei zwei Mutanten, giant und lethal-giant-larvae. *Chromosoma (Berl.)*, **10**, 654–678.

Beermann, W. (1952). Chromomerenkonstanz und spezifische Modifikationen der Chromosomenstruktur in der Entwicklung und organdifferenziening von *C. tentans*. *Chromosoma (Berl.)*, **5**, 139–198.

Beermann, W. (1972). Chromomeres and genes. In *Developmental Studies on Giant Chromosomes* (W. Beermann, ed.), Springer Verlag, Berlin, Heidelberg, and New York, pp. 1–33.

Berendes, H. D., and Ashburner, M. (1978). The salivary glands. In *The Genetics and Biology of Drosophila*, Vol. 2b. (M. Ashburner and T. R. F. Wright, eds), Academic Press, London and New York, pp. 453–498.

Bridges, C. B. (1935). Salivary chromosome maps. *J. Hered.*, **26**, 60–64.

Bridges, C. B. (1936). The Bar 'gene' duplication. *Science*, **83**, 210–211.

Chalkley, R., and Hunter, C. (1975). Histone–histone propinquity by aldehyde fixation of chromatin. *Proc. Natl Acad. Sci.*, **72**, 1304–1308.

Cohen, L. H., and Gotchel, B. V. (1971). Histones of polytene and nonpolytene nuclei of *Drosophila melanogaster*. *J. Biol. Chem.*, **246**, 1841–1848.

Coluzzi, M. (1968). Cromosomi politenici delle cellule nutrici ovariche nel complesso *gambiae* del genre *Anopheles*. *Parassitologia*, **10**, 179–183.

Coluzzi, M., Sabatini, A., Petrarca, V., and DiDeco, M. A. (1979). Chromosomal differentiation and adaptation to human environments in the *Anopheles gambiae* complex. *Trans. Roy. Soc. Trop. Med. Hyg.*, **73**, 483–497.

Curtis, C. F. (1980). Translocations, hybrid sterility, and the introduction into pest populations of genes favorable to man. In *Genetics in Relation to Insect Management* (M. A. Hoy and J. J. McKelvey, eds), The Rockefeller Foundation, New York, pp. 19–30.

Daneholt, B., and Edström, J.-E. (1967). The content of DNA in individual polytene chromosomes of *C. tentans*. *Cytogenetics*, **6**, 350–356.

Dick, C., and Johns, E. W. (1968). The effect of two acetic acid fixatives on the histone content of calf thymus deoxyribonucleoprotein and calf thymus tissue. *Exp. Cell Res.*, **51**, 626–632.

Edström, J.-E., Rylander, L., and Francke, C. (1980). Concomitant induction of a Balbiani ring and a giant secretory protein in *Chironomus* salivary glands. *Chromosoma (Berl.)*, **81**, 115–124.

French, W. L., Baker, R. H., and Kitzmiller, J. B. (1962). Preparation of mosquito chromosomes. *Mosquito News*, **22**, 377–383.

Gall, J. G., Cohen, E. H., and Polan, M. L. (1971). Repetitive DNA sequences in *Drosophila*. *Chromosoma (Berl.)*, **33**, 319–344.

Heitz, E., and Bauer, H. (1933). Beweise für die Chromosomennatur der Kernschleifen in den Knavelkernen von *Bibio hortulans*. *Z. Zellforsch. Mikrosk. Anat.*, **17**, 67–82.

Heitz, E. (1934a). Die somatische Heteropyknose bei *Drosophila melanogaster* und ihre genetische Bedeutung. *Z. Zellforsch.*, **20**, 237–287.

Heitz, E. (1934b). Über α- und β-heterochromatin sowie Konstanz und Bau der Chromomeren bei *Drosophila*. *Biol. Zbl.*, **54**, 588–609.

Judd, B. H., Shen, M. W., and Kaufman, T. C. (1972). The anatomy and function of a segment of the X chromosome of *D. melanogaster*. *Genetics*, **71**, 139–156.

Keyl, H. G. (1965). A demonstrable local and geometric increase in the chromosomal DNA of *Chironomus*. *Experientia (Basel)*, **21**, 191–199.

Lefevre, G. (1974). The relationship between genes and polytene chromosome bands. *Annu. Rev. Genet.*, **8**, 51–62.

Lewis, E. B. (1960). A new standard food medium. *Drosophila Inf. Serv.*, **34**, 117–118.

McKenzie, S. L., Henikoff, S., and Meselson, M. (1975). Localization of RNA from heat induced polysomes at puff sites in *D. melanogaster*. *Proc. Natl Acad. Sci.*, **72**, 1117–1121.

Painter, T. S. (1933). A new method for the study of chromosome rearrangements and the plotting of chromosome maps. *Science*, **78**, 585–586.

Painter, T. S. (1934). Salivary chromosomes and the attack on the gene. *J. Hered.*, **25**, 465–476.

Pelling, C. (1965). Ribonuklein saure Synthese der Reisenchromosomen. *Chromosoma (Berl.)*, **15**, 71–122.

Plagens, U., Greenleaf, A. L., and Bautz, E. K. F. (1976). Distribution of RNA polymerase on *Drosophila* polytene chromosomes as studied by indirect immunofluorescence. *Chromosoma (Berl.)*, **59**, 157–165.

Rudkin, G. T. (1961). Cytochemistry in the ultraviolet. *Microchem. J. Symp. Ser.*, **1**, 261–276.

Rudkin, G. T. (1965). Relative mutabilities of DNA in regions of X chromosome of *Drosophila melanogaster*. *Genetics*, **52**, 665–681.

Rudkin, G. T. (1969). Non replicating DNA in *Drosophila*. Genetics (Suppl.) **61**, 227–238.

Rudkin, G. T. (1972). Replication in polytene chromosomes. In *Developmental Studies on Giant Chromosomes* (W. Beermann, ed.), Springer Verlag, Berlin, Heidelberg, and New York, pp. 59–85.

Rylander, L., and Edström, J.-E. (1980). Large sized nascent protein as dominating component during protein synthesis in *Chironomus* salivary glands. *Chromosoma (Berl.)*, **81**, 85–99.

Rylander, L., Pigon, A., and Edström, J.-E. (1980). Sequences translated by Balbiani ring 75 s RNA *in vitro* are present in giant secretory protein from *Chironomus tentans*. *Chromosoma (Berl.)*, **81**, 101–113.

Silver, L. M., and Elgin, S. C. R. (1976). A method for determination of the *in situ* distribution of chromosomal proteins. *Proc. Natl Acad. Sci.*, **73**, 423–427.

Silver, L. M., and Elgin, S. C. R. (1978). Immunological analysis of protein distributions in *Drosophila* polytene chromosomes. In *The Cell Nucleus*, Vol. 5 (H. Busch, ed.), Academic Press, New York and London, pp. 215–262.

Silver, L. M., Wu, C. E. C., and Elgin, S. C. R. (1978). Immunofluorescent techniques in the analysis of chromosomal proteins. In *Methods in Cell Biology* (G. Stein, J. Stein, and L. Kleinsmith, eds), Academic Press, New York and London, pp. 151–167.

White, G. B. (1981a). Academic and applied aspects of mosquito cytogenetics. In *Insect Cytogenetics* (R. L. Blackman, G. M. Hewitt, and M. Ashburner, eds), Symposia of the Royal Entomological Society of London, No. X, Blackwell Scientific Publications, Oxford, pp. 245–274.

White, G. B. (1981b). Mosquito cytogenetics: a classified bibliography. *Mosquito System.*, **13**, in press.

Yoon, J. S., Richards, R. H., and Wheeler, M. R. (1973). Technique for improving salivary chromosome preparations. *Experientia (Basel)*, **29**, 639–641.

Zegarelli-Schmidt, E. C., and Goodman, R. (1981). The Diptera as a model system in cell and molecular biology. *Internat. Rev. Cytol.*, **71**, 245–363.

References to polytene chromosome maps of some widely studied *Drosophila* species

D. ananassae

Hinton, C. W., and Downes, J. E. (1975). The mitotic, polytene and meiotic chromosomes of *Drosophila ananassae*. *J. Hered.*, **66**, 353–361.

Moriwaki, D. and Ito, S. (1969). Studies on puffing in the salivary gland chromosomes of *Drosophila ananassae*. *Jap. J. Genet.*, **44**, 129–138.

D. hydei

Berendes, H. D. (1963). The salivary gland chromosomes of *Drosophila hydei* Sturtevant. *Chromosoma (Berl.)*, **14**, 195–206.

D. lebanonensis

Berendes, H. D. and Thijssen, W. T. M. (1971). Developmental changes in genome activity in *Drosophila lebanonensis casteeli* Pipkin. *Chromosoma (Berl.)*, **33**, 345–360.

D. melanica

Stalker, H. D. (1966). The phylogenetic relationships of the species in the *Drosophila melanica* group. *Genetics*, **53**, 327–342.

Ward, C. L. (1952). Chromosome variation in *Drosophila melanica*. *Univ. Texas Publ.*, no. 5204, 137–158.

D. melanogaster

Bridges, C. B. (1935). Salivary chromosome maps. *J. Hered.*, **26**, 60–64.
Bridges, C. B. (1938). A revised map of the salivary gland X chromosome of *Drosophila melanogaster*. *J. Hered.*, **29**, 11–13.
Bridges, C. B., and Bridges, P. N. (1939). A new map of the second chromosome: a revised map of the right limb of the second chromosome in *Drosophila melanogaster*, *J. Hered.*, **30**, 475–476.
Bridges, P. N. (1941a). A revised map of the left limb of the third chromosome of *Drosophila melanogaster*. *J. Hered.*, **32**, 64–65.
Bridges, P. N. (1941b). A revision of the salivary gland 3R-chromosome map of *Drosophila melanogaster*. *J. Hered.*, **32**, 299–300.
Bridges, P. N. (1942). A new map of the salivary gland 2L-chromosome of *Drosophila melanogaster*. *J. Hered.*, **33**, 403–408.
Lefevre, G., Jr (1976). A photographic representation and interpretation of the polytene chromosomes of *Drosophila melanogaster* salivary glands. In *The Genetics and Biology of Drosophila*, Vol.1A (M. Ashburner and E. Novitski, eds), Academic Press, New York and London, pp. 31–66.

D. pseudoobscura

Stocker, A. J., and Kastritsis, C. D. (1972). Developmental studies in *Drosophila*. III. The puffing patterns of salivary gland chromosomes of *Drosophila pseudoobscura*. *Chromosoma (Berl.)*, **37**, 139–176.

D. subobscura

Loukas, M., Krimbas, C. B., Mavragani-Tsipidou, P., and Kastritsis, C. D. (1979). Genetics of *Drosophila subobscura* populations. VIII. Allozyme loci and their chromosome maps. *J. Hered.*, **70**, 17–26.

D. virilis

Hsu, T. C. (1952). Chromosomal variation and evolution in the *virilis* group of *Drosophila*. *Univ. Texas Publ.*, no. 5204, 35–72.

6

Lampbrush chromosomes

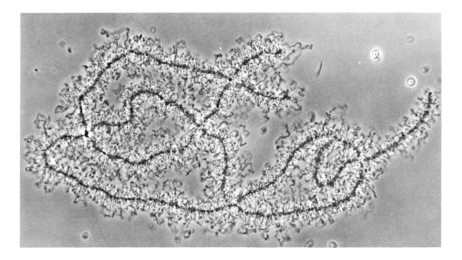

A phase contrast micrograph of a freshly isolated lampbrush bivalent from *Notophthalmus viridescens*. [Reproduced with permission from Gall, J., in *Methods in Cell Physiology*, Vol. 2 (D. Prescott, ed), Academic Press, New York, pp. 37–60 (1966). © 1966 Academic Press Inc.]

The largest and most spectacular form of chromosome is found in the growing oocytes of most animals and has been most extensively studied in oocytes of amphibians. These so-called 'lampbrush chromosomes' are in many ways one of the most favourable of all for exprimental manipulation, and at various times right up to the present day they have served as foci for attacks on a number of absolutely fundamental problems of chromosome organization and activity.

In the context of modern high resolution studies of genetic structure, function, and evolution, lampbrushes offer unique opportunities for visualizing genes in action. Nowhere else, except in polytene chromosomes, is it possible actually to see and interpret the direct consequences of events that are known to be taking place at the molecular level and that involve specific DNA sequences. But the real beauty of lampbrush chromosomes is that they can be prepared for microscopy, examined, and researched with the minimum of expense and equipment. Even in these days of high technology, lampbrushes present us with a wide range of fascinating cytogenetic problems that can be tackled by anyone who is reasonably good with their hands and with a microscope.

Lampbrush chromosomes were first seen in sections of axolotl oocytes by Flemming in 1882 and then again 10 years later by Ruckert (1892) in dogfish oocytes. The name lampbrush comes from Ruckert who likened the objects to the 19th century lampbrush, equivalent to the 20th century test tube brush. The lampbrush type of chromosome is now known to be characteristic of growing oocytes in the ovaries of almost all animals, with certain notable exceptions. The chromosomes are in a highly extended form, sometimes reaching more than a millimetre in length. They are essentially bivalents that persist through the extended diplotene phase of the first meiotic division, a phase that can last as long as 6 months or more in certain organisms. In the biological sense, these chromosomes are odd only in so far as they are confined to meiotic prophase in germ cells, but they are a general phenomenon in so far as they are found in nearly all animals.

Lampbrush chromosomes have recently been the subject of a number of useful reviews. Two 'state of the art' reviews were published by Macgregor in 1977 and then again in 1980. A wide-ranging review with a strong biochemical and molecular emphasis was published by Sommerville in 1977; H. G. Callan's Croonian Lecture to the Royal Society in 1981 (Callan, 1982) places lampbrushes in good perspective with a strong historical emphasis.

The essence of the lampbrush technique is the free-hand manipulation of animal oocytes and their nuclei. Growing oocytes in the ovaries of most animals other than mammals are relatively large cells with diameters ranging from a few tens of micrometres up to several millimetres. For the most part they can be seen with the naked eye and easily manipulated with fine forceps and the help of a low power dissecting binocular microscope. Oocyte nuclei are also large, those from amphibians sometimes reaching diameters of more than 1 mm. So they too can be seen and manipulated with relative ease. The lampbrush chromosomes that are found inside these oocyte nuclei are exceedingly delicate structures and no further progress beyond the pioneering studies of Flemming and Ruckert was possible until techniques could be devised for dissecting the chromosomes out of their nuclei and examining them in a life-like condition, separated from the remainder of the nuclear contents. The first steps towards such a technique were taken by Duryee in 1937. The technique that is described here was introduced by Gall in 1954. Students of these chromosomes may find it both interesting and useful to look back at the early works of Ruckert, Duryee, and Gall and examine the progress that these men made in relation to the facilities and scientific background that existed in their respective times. In the procedure normally used today an oocyte is punctured with a needle, the nucleus gently squeezed out of the hole, picked up with a fine Pasteur pipette, and transferred to saline in a chamber constructed by boring a hole through a microscope slide and sealing a coverglass across the bottom of the hole with wax. The nuclear membrane is then removed and the nuclear contents, including the lampbrush chromosomes, disperse and come to lie flat on the upper surface of the coverglass that forms the floor of the chamber. Then, by using a phase contrast microscope with an inverted optical system, essentially looking up through the bottom of the chamber, the chromosomes can be examined in a fresh and unfixed condition with the highest resolution obtainable with a light microscope.

I. Choice of animals

It is probably possible to get the chromosomes out of oocytes from almost any animal using basic lampbrush technology. The critical limitations are the accessibility and size of the oocytes. The would-be investigator should not be deterred from 'having a go' at material from any animal after first working through the basic technique with one of the large and straightforward systems. Lampbrushes have been isolated from many amphibians, a number of reptiles, worms, molluscs, and insects. Curiously, and undoubtedly most significantly, they have also been seen in the giant nucleus of the unicellular alga *Acetabularia*. The initial approach to all these systems has been the same: get the nucleus out, transfer it to an observation chamber, take off its membrane, and look at its dispersed contents with an inverted phase contrast microscope.

Newts are certainly the best and most convenient source of material for lampbrush studies. They are easy to obtain, easy to keep in the laboratory, and they survive minor operations to remove small pieces of their ovary without killing the animal. They offer a good range of oocyte sizes throughout the greater part of the year. They have large genomes and therefore large chromosomes, and in some of the commonest species the lampbrush chromosomes show a range of remarkable linear differentiations that have proved especially useful for chromosome identification and characterization. It is a curious and fortunate fact that on top of this impressive list of attributes, the oocyte nuclei and lampbrush chromosomes of newts just happen to be exceptionally easy to handle and aesthetically pleasing to look at. Had lampbrush studies begun and been developed with *Xenopus laevis* or with any species of *Rana*, for example, or with plethodontid salamanders, then progress would have been disappointing and slow, and it would probably not have been worth writing this chapter today.

All species of newt belonging to the genera *Triturus*, *Notophthalmus*, and *Taricha* are entirely suitable for lampbrush studies. The two animals that are most commonly used are *Triturus cristatus carnifex* from Italy and *Notophthalmus viridescens* from the United States. Most species of newt can quite easily be collected from ponds in the early spring during May or June. Locations of ponds that have good newt populations can usually be obtained from local naturalists in schools, universities, or museums provided that some assurance is given that populations will not be over-exploited. Nearly all species of *Triturus*, *N. viridescens*, and *Taricha granulosa* are obtainable through suppliers listed in Appendix 1, provided that orders are placed well in advance. Suppliers usually receive one main shipment of animals per year, most commonly in April or May. Therefore the best strategy is to place a provisional order sometime during the previous winter to be assured of obtaining animals the following spring. The leading suppliers in the United States and the UK now maintain year-round supplies of *T. c. carnifex* and *N. viridescens*.

Would-be 'lampbrushers' who are operating on a limited budget and require small numbers of animals mainly for teaching purposes are advised to collect their own and make do with whatever species of newt are locally available. Those who specifically wish to use crested newts or *N. viridescens* in their teaching or research are advised to obtain the animals from accredited suppliers. It seems likely that supplies of all species of newt may soon become limited as European and United States governments introduce legislation for protection of what may well be endangered species. Licences are already required for import and export of these animals and are also needed for collection of certain species. In this sense it is undoubtedly faster, less complicated, and less expensive in terms of both time and money to use the good services of properly licensed and organized suppliers. Advice concerning British laws governing collection, export and import of live animals to and from the UK may be obtained by writing to the Department of The Environment, Wildlife Conservation Licensing Section, Tollgate House, Houlton Street, Bristol BS2 9DJ.

A. Notes on individual species

Triturus cristatus carnifex is by far the best animal for lampbrush studies. It is easy to obtain in spring and early summer and presently abundant in the wild. It is easy to ship, easy to keep in captivity for long periods, and it breeds successfully in captivity. It survives laparotomy well for removal of ovary (Home Office Licence needed in the UK). Its oocyte nuclei are easy to isolate and manipulate, its lampbrush chromosomes are very well characterized and have lots of interesting features, and they are the subject of an extensive range of published work. For teaching purposes, one good female newt will provide enough material for one useful session on lampbrush technology with five students.

T. c. cristatus is as good in every respect as *T. c. carnifex* but is less abundant, less readily obtainable from suppliers, and its oocyte nuclei and chromosomes are a little more difficult to handle successfully.

T. marmoratus is likewise as good as *T. c. carnifex* but is quite hard to obtain from suppliers and is usually expensive, despite being abundant and widespread in Europe. It is a large and handsome newt with a large ovary. Its oocyte nuclei and lampbrush chromosomes are easy to handle, and the chromosomes have many interesting features.

T. alpestris is an abundant and useful newt for lampbrush studies. It is common in south-eastern England, although this is not shown on published distribution maps, as well as being widespread on continental Europe. It is a smaller newt and may not survive laparotomy. Its oocyte nuclei and chromosomes are relatively easy to handle.

T. vulgaris and *T. helveticus* are both entirely suitable for lampbrush studies. They are abundant throughout Europe and easy to obtain from suppliers. They are small newts that usually have to be killed to obtain oocytes. The oocyte nuclei and chromosomes are easy to handle.

Notophthalmus viridescens is the favourite animal for lampbrush studies in the United States. It is abundant in the wild at certain times of year; it is usually easy to obtain from suppliers in the USA and other parts of the world; it is easy to ship; it will not usually feed in captivity but can be kept for long periods at 4–5°C. It is a small newt that survives laparotomy if very carefully handled, but often has to be killed to obtain sufficient oocytes. The oocyte nuclei and chromosomes are easy to handle. The chromosomes have many interesting features and have been well characterized.

Taricha granulosa is an abundant newt in the western United States and is usually obtainable from suppliers in the USA and the UK. It is a large newt that can easily be laparotomized successfully, and during the autumn and winter it has many oocytes that are suitable for lampbrush studies. Its oocyte nuclei and chromosomes are easy to handle, although its oocytes will not remain in good condition for lampbrushology for more than 1–2 h after they have been removed from the animal.

A list of urodele species that have been investigated and are known to be suitable for lampbrush studies without resort to special methods, other than minor adjustments in the salines used for chromosome isolation, is given in Table 6.1 together with some other species for which special methods have been worked out and published. At the end of this chapter some comments are offered concerning the problems that investigators are likely to encounter and how to overcome them when dealing with other animals such as salamanders, frogs, toads, or reptiles.

Table 6.1   Species known to be useful for lampbrush chromosome studies

Species	Reference
Triturus cristatus	Callan and Lloyd, 1960
T. marmoratus	Nardi *et al.*, 1972
T. alpestris	Mancino *et al.*, 1972
T. vulgaris	Barsacchi *et al.*, 1970
T. helveticus	Mancino and Barsacchi, 1966
T. italicus	Mancino and Barsacchi, 1969
Salamandra salamandra	Mancino *et al.*, 1969
Pleurodeles waltlii	Lacroix, 1968
P. poireti	Lacroix, 1968
Notophthalmus viridescens	Gall, 1954; Callan and Lloyd, 1975
Ambystoma mexicanum	Callan, 1966
Taricha granulosa	Kezer *et al.*, 1982
Plethodon cinereus	Vlad and Macgregor, 1975
Xenopus laevis	Jamrich and Warrior, 1982
Ascaphus truei	Macgregor and Kezer, 1970
Bipes biporus	Macgregor and Klosterman, 1978

II.   Maintenance of animals

All species of *Triturus*, *Notophthalmus*, and *Pleurodeles* should be kept in 5–10 cm of water and provided with a rough stone placed in such a way that the animals can climb out of the water. The stone is also essential to assist newts to rid themselves of moulted skin. These animals can be fed on *Tubifex* worms, earthworms (not brandling worms, *Eisenia foetida*, which are toxic and will not be taken by newts), or chopped liver. If *Tubifex* are used then the water should be lightly aerated, running, or be changed at least every 2 days. If earthworms or liver are used then the animals should be hand-fed and the uneaten material should be removed from the tank immediately. Temperature is best maintained at around 16°C with a constant day/night lighting cycle. Tanks should be securely covered. Newts can climb!

Terrestrial salamanders should be placed in closed tanks or jars with damp paper towelling. The towelling should be of a coarse fibrous kind, not the soft toilet variety. It should be thoroughly moistened but there should be no free liquid in the jar. Larger salamanders are best fed by hand on earthworms. Smaller animals, such as plethodontid salamanders, can most easily be maintained on live *Drosophila*.

III. Obtaining oocytes and chromosomes

The protocols given in this section apply specifically to newts of the genus *Triturus*. Any special modifications of technique that are required for other species will be mentioned at appropriate points.

A. Obtaining oocytes by laparotomy with anaesthesia

(i) Equipment needed

(a) Ice bucket with crushed ice
(b) Small square flat-topped embryo cup or watchglass with a glass cover
(c) Vaseline or paraffin grease
(d) Paper towelling
(e) Very good fine pointed scissors
(f) Two pairs of clean fine pointed (but not sharp) watchmakers' forceps
(g) 5–10 cm of 3/0 catgut suture threaded through a number 1 surgical needle (eye half curved) and placed to soak in distilled water
(h) A low power long working distance dissecting microscope
(i) 0.1% solution of MS222 (ethyl *m*-aminobenzoate methanesulphonic acid salt, or tricaine methane sulphonate; Sigma cat. no. E1626), enough to fill a large Petri dish so as to almost submerge the animal to be anaesthetized (note that MS222 is expensive, so be sparing with it) *or* a glass jar of about 1 l capacity with a well fitting top and a tight pad of cotton wool stuck securely to the inside of the top and soaked in ether
(j) A glass jar of about 1 l capacity with a well fitting top (for recovery)
(k) Surgical gloves of the disposable variety

(ii)   Procedure

Anaesthetize a newt in MS222. This should take between 5 and 15 min and is judged to be adequate when the animal has ceased all voluntary movement. If MS222 is not available then light ether anaesthesia is entirely adequate, although this should be for the shortest possible time and the animal should be exposed to the ether vapour only. Contact with ether liquid will kill the animal.

Wash the newt thoroughly in tap water after anaesthesia to remove all traces of anaesthetic and slime. Note that newt slime is a powerful histamine releaser and should be treated with great respect by those who are prone to allergic reactions. Wear gloves and wash everything thoroughly after use.

Place the animal on its back on a piece of wet paper towelling and make a small longitudinal incision posteriorly in the ventral body wall well to the left or right of the mid line. The cut need not be more than about 5 mm in length. Cut first through the skin and underlying muscle layer and then very carefully through the thin peritoneum. Take special care not to cut deeply as the inflated lungs often lie directly beneath the peritoneum and damage to the lung will assuredly kill the newt.

Deflect the viscera inside the cut, grasp the ovary which is usually quite clearly visible at this stage and draw it out through the cut. Snip off a piece of ovary, taking just as much material as you think you are likely to need. Try hard at this stage not to cut through any major blood vessels. Some investigators like to carry out most of the laparotomy with the help of a low power dissecting binocular microscope. This is a matter of personal choice, but for the less experienced investigator it may reduce the chance of doing serious damage to the animal.

Place the piece of ovary immediately into a clean dry embryo cup or watchglass, seal the top of the container with vaseline and place it on ice. Do not freeze it. Kept in this way the ovary will remain in good condition for lampbrush work for up to 24 h, although there will be some changes in the characteristics of the oocyte nuclei and the appearance of the lampbrush chromosomes in this time.

Meanwhile gently replace the remaining attached ovary in the newt's body cavity and sew up the wound with thoroughly wetted and softened catgut suture. Synthetic sutures are not usually successful but can be resorted to if catgut is definitely not available. When sewing the wound take care to pass the needle through the peritoneum as well as the body wall. Do not pull the stitches too tight. Use reef (square) knots, leaving generous ends of about 2 mm on either side of the knot. Two or at the most three stitches should be sufficient.

Finally wash the newt thoroughly and place it in a jar with 2–3 cm of clean water to recover. Recovery should take 15 min to 2 h.

B. Making lampbrush preparations

(i) Equipment needed

(a) A binocular dissecting microscope. This should be of a simple kind with a fixed magnification of around 20× or a variable 'zoom' magnification of between 10× and 40×. The microscope should be fitted with a calibrated micrometre eyepiece and a black stage plate. It should have an associated incident light source that can be focused to a sharp, intense beam shining on to the middle of the stage at a low angle. Possibly the best dissecting microscopes for lampbrush work are the Carl Zeiss Stereozoom III and IV and the Nikon SMZ 10. Above all, the microscope should be simple to operate and it should be placed on a bench where there is plenty of light and elbow room and with a chair that is just the right height for working with that particular microscope at that particular bench. Comfort and stability are of the utmost importance when carrying out fine scale dissections.

(b) A good quality mounted dissecting needle, preferably of stainless steel, and with a moderately sharp but slightly rounded point.

(c) Two pairs of number 4 watchmakers' forceps.

(d) One or two pairs of number 5 watchmakers' forceps. These will be used exclusively for handling germinal vesicles and must therefore have very fine points indeed. They should not be of stainless steel since this cannot be honed down to a fine and strong enough point. The tips of the forceps should be honed and polished to the finest possible points that meet exactly. The best materials for sharpening forceps are a small, well oiled Arkansas stone and a piece of the finest available waterproof sandpaper. Examine the points of the forceps under the dissecting binocular and judge whether their points are good enough for working under approximately 30× magnification.

(e) Tungsten wire needles. Tungsten wire can be purchased from most suppliers of equipment for electron microscopy. The best thickness is about 0.35–0.40 mm. Fuse a piece of wire about 3 cm long into the end of a piece of Pyrex glass tubing about the size of a pencil. Leave 2 cm of the wire projecting from the end of the holder. To point the wire, take a nickel crucible half-filled with sodium nitrite and stand it in a pipe-clay triangle over a hot bunsen flame. Melt the sodium nitrite completely and then dip the needle into it. *Take care*, hot sodium nitrite 'spits' and can be dangerous. If the nitrite is not hot enough then the wire will merely be etched clean. If the nitrite is hot enough then the wire will immediately become incandescent. Dip the tip in and out several times then wash it under a tap and look at it with the dissecting binocular. The point should be very sharp and very clean. The sharper the better. Wash the needle thoroughly, then flame it a few times in the bunsen. It is now ready for use. Put it somewhere safe immediately.

(f) Narrow bore Pasteur pipette. These are best prepared from long form disposable Pasteur pipettes. Hold the stem of the pipette over a hot narrow bunsen flame at the position indicated on Figure 6.1, rotating it continuously with both hands. When the heated region is completely flexible, pull it out with a swift but smooth motion to a distance of about 1 m. This may take a little practice. Break off the new pipette leaving a long stem. Square off the end of the pipette at about 5 cm from the shoulder by stroking it with a diamond pencil and then lightly snapping it off. This should produce a completely square pipette tip of about 0.5 mm internal diameter. Touch the end of the pipette very briefly to the side of a hot bunsen flame to round off the sharp edges. Only the lightest touch will suffice. Inspect the pipette tip under the dissecting binocular to see that it is of the right size, and has a smooth square end. It is useful to make a stockpile of about 20 pipettes at one session.

(g) Observation chambers. There are two ways in which these can be made. The first involves microscope slides with 5 mm holes bored through

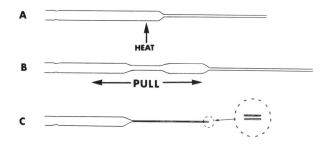

Figure 6.1   Diagrams to show the two stages in the construction of a 'lampbrush' pipette from a long-form disposable glass Pasteur pipette. Figure 6.1(A) shows the position (heat) at which the disposable pipette should be heated prior to drawing out to form the long necked narrow mouthed pipette suitable for handling germinal vesicles. Figure 6.1(C) shows the shape and the approximate proportions of the finished product

them. These can usually be made quite easily by a skilled glassworker using an ultrasonic or diamond drill and boring through stacks of 10 or more slides at a time. The second way involves the use of ordinary microscope slides and small plastic rings that can be constructed by anyone with a sheet of suitable plastic, a paper punch and some suitable adhesive. Bored slides are essential for good high resolution work with an inverted phase contrast microscope. Slides with plastic rings are entirely adequate for low power work and for making permanent preparations for examination with an ordinary microscope. The methods for making up each type of chamber are described separately below.

To prepare chambers with bored slides, place a number of these next to one another on a clean dark background. With a heated copper or brass rod 1–2 mm thick put two small dabs of paraffin wax on either side of the hole, each about 1–2 mm out from the edge of the hole. Thoroughly clean some coverglasses 18 or 22 mm square (number 1½) from 95% ethanol and place one over the hole of each bored slide, sitting squarely on the two dabs of wax. Now wave a small gas flame over each coverglass until the wax below has melted and flowed evenly round to form a complete seal between the slide and the coverglass. A moderate flame from a small bunsen burner is entirely suitable for this job, but it must be kept moving, otherwise the coverglass and/or the slide will crack. Take care not to use too much wax; if there is too much then it will ride up the walls of the hole in the slide and it will be impossible to fill the chambers with aqueous liquid thereafter. It is best to have at least ten chambers available. Clean coverglasses 18 mm and 22 mm square will be needed later as covers for the chambers. Have these to hand.

A scale diagram of a lampbrush chamber made from a bored slide is shown in Figure 6.2. Chambers such as these are designed to allow the chromosomes to be spread out over the base of the chamber and examined in a fresh and unfixed condition with dry and oil-immersion objectives fitted to an inverted phase contrast microscope. If the investigator wishes to make permanent preparations, then the chambers should be constructed with a normal microscope slide forming the base of the chamber. The procedure is no different except that more wax and a little more heat are required to seal the two slides together. When constructing chambers of this kind make absolutely sure that the two slides are in perfect alignment with one another. The use of chambers of this kind will be explained later.

The construction of chambers with plastic card requires ordinary clean microscope slides, some double-sided Sellotape or Scotch tape, some pieces of Teflon (or equivalent) sheeting about 0.5 mm thick (Slaters Plasticard Ltd), one or two coverglasses 22 mm square, a pair of scissors, and an

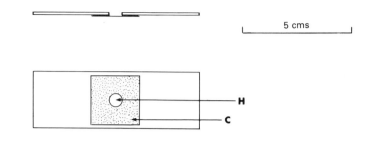

Figure 6.2   Scale drawing in section and plan of a lampbrush observation
chamber constructed from a bored slide and a coverglass (C) sealed over
the hole (H) with wax

ordinary office machine for punching holes in pieces of paper that are to be
filed in ring folders. First cut a piece of double-sided Sellotape about 30 mm
square; the exact dimensions are unimportant except that it should be more
than 22 mm square. Cut a piece of plastic sheeting to about the same size.
Stick the Sellotape firmly and evenly to one side of the piece of plastic
sheeting. Trim both with scissors to 22 mm square using the 22 mm square
coverglass as a template. Place the plastic–Sellotape square in the punch and
make a hole as nearly as possible in the middle of it. The hole must be clean
and round. If it is not, then get a new punch or have the existing one
sharpened. Remove the backing from the double-sided Sellotape on the
square of plastic card. Stick the square of plastic card firmly and evenly on
to the middle of a clean microscope slide. The chamber is now complete and
ready for use. Preparations made in such chambers should be covered with
coverglasses that are smaller than the plastic rectangle.

Chambers made with plastic card have the advantage that they are easy to
construct, they can be used with a normal microscope employing objectives
up to 25× dry magnification, and they can be used to make permanent
preparations. Their main disadvantage is that they never allow uncompro-
misingly good microscopy since the depth of the chamber affects to some
extent the performance of the objective, and they can never be used with
oil-immersion objectives since these will always have working distances that
are shorter than the depth of the chamber. Nevertheless, for the investigator
who cannot obtain bored slides or who simply cannot justify the expense and
trouble of making them, plastic card chambers offer a real and entirely
practicable alternative.

(h) Moist chambers. These are needed for even short term storage of
newly made lampbrush preparations. They are designed to prevent preparations from drying out while the chromosomes and nucleoplasm are dispersing
immediately after removal of the nuclear membrane. They are also useful
for keeping fresh preparations while the investigator takes time to examine,

photograph, or draw them. The simplest moist chamber consists of a 10 cm square Petri dish containing a 9 cm diameter filter paper, about 10 ml of the same saline into which the lampbrush chromosomes were dissected, and a piece of glass rod bent into a U in such a way as to form a supporting bridge to keep the lampbrush chambers clear of the wet filter paper.

(i) Microscopes. Lampbrush chromosomes lying at the bottom of any of the observation chambers described in this chapter can be seen with any good phase contrast microscope at objective magnifications up to 16× with bored slide chambers and up to 25× with plastic card chambers. However, for good phase contrast microscopy an inverted microscope is essential. Good inverted microscopes for lampbrush work are currently obtainable from Carl Zeiss and Nikon. An electronic flash is desirable but not essential for photomicrography of lampbrushes because the chromosomes show extensive Brownian motion and this usually precludes high resolution photomicrography with a normal light source.

(ii)   Chemical media

A medium that is ideal for the study of unfixed lampbrush chromosomes should have the following properties. After isolation of an oocyte nucleus and removal of its membrane the nucleoplasm should disperse over the bottom of the observation chamber within 10–20 min, allowing the chromosomes to fall apart from one another and spread out over the base of the chamber. The ribonucleoprotein of the lateral loops of the chromosomes should remain in position in the loops; it should not stiffen or coagulate; and it should not show signs of swelling or hydration. It should be possible to keep the chromosomes in good condition without change in their appearance for periods of up to 3–4 hours at 20°C.

For some species of amphibian these conditions are easy to meet. For others they can only be met with difficulty and by accepting a range of compromises. In general, the newts are 'easy' animals. All other animals require more or less fiddling with isolation media in order to achieve good results.

The basic medium in use for lampbrush work is an unbuffered 5 : 1 mixture of 0.1 M KCl and 0.1 M NaCl (5 : 1 K/NaCl) with a pH in the range 6–7.5. This medium is always used for the removal of the germinal vesicle nucleus from the oocyte no matter what species is being studied. More often than not the nucleoplasm will not disperse in 5 : 1 K/NaCl but remains as a stiff, rounded mass of jelly after removal of the nuclear membrane, so preventing dispersal of the chromosomes. In this case $CaCl_2$ should be added to a concentration of not more than $\sim 10^{-4}$ M, and this will usually help to solubilize the nucleoplasm and encourage dispersal of the chromosomes. Aim at the lowest effective concentration of $CaCl_2$ in the dispersion medium since the more slowly the nucleoplasm solubilizes the easier it is to remove the nuclear membrane without damaging the chromosomes. Note that $CaCl_2$ should not be added to the medium in which the nucleus is removed from the oocyte, but only to the medium that is placed in the observation chamber where the nuclear membrane is removed. In what follows the medium used for removing the nucleus from the oocyte will be called the 'isolation medium', and that used in the observation chamber will be called the 'dispersion medium'.

Many species have oocyte nucleoplasm that is hard to disperse. Various options are available for overcoming this problem. The dispersion medium may be diluted, although this will certainly result in loss of chromosome material if the concentration is reduced to less than about 0.05 M. The pH of the dispersion medium can be adjusted, but only within the range 6–8. A trace of formaldehyde or cytochalasin-B (5μg/ml) can be added to the dispersion medium, or the preparations, originally made in 5 : 1 K/NaCl, can be exposed for a few minutes to formaldehyde vapour. Some useful information on the kinds of combinations of tricks that have been applied to the dispersal problem can be found in papers by Callan (1966), who was working with axolotl chromosomes, Macgregor and Kezer (1970), working with lampbrushes from the Pacific tailed frog (*Ascaphus truei*), and Vlad and Macgregor (1975), working with chromosomes from plethodontid salamanders. Details of the dispersion media used by these authors are given in Table 6.2.

Table 6.2   Isolation and dispersion media used by various authors in the preparation of lampbrush chromosomes

Species	Isolation medium	Dispersion medium	Reference
Triturus cristatus	0.1 M 5 : 1 K/NaCl	0.1 M 5 : 1 K/NaCl + 10^{-4} M $CaCl_2$	Gall and Callan, 1962
Ambystoma mexicanum	0.1 M 5 : 1 K/NaCl	0.06 M 5 : 1 K/NaCl + 0.3×10^{-4} M $CaCl_2$	Callan, 1966
Plethodon cinereus	0.05 M 5 : 1 K/NaCl	0.05 M 5 : 1 K/NaCl, 7 parts; 0.006 M phosphate buffer, pH 7, 3 parts; formaldehyde, 0.5%	Vlad and Macgregor, 1975
Ascaphus truei	0.1 M 5 : 1 K/NaCl	0.05 M 5 : 1 K/NaCl, 3 parts; 0.001 M KH_2PO_4, 7 parts; formaldehyde, 0.5%	Macgregor and Kezer, 1970

Formaldehyde is made up by dissolving 20 g paraformaldehyde powder in water, making up to a final volume of 100 ml, raising the pH to 10 with NaOH, heating to between 60° and 80°C until the powder is completely dissolved, cooling, filtering and neutralizing with HCl.

Lampbrush media are best made up in large batches of 1 l or more and then dispensed into 150 ml Erlenmeyer flasks for the nuclear isolation medium and 50 ml flasks for the dispersion medium. The flasks should be capped with foil and autoclaved. If no autoclave is available then boiling will suffice. Media that have been autoclaved can be kept indefinitely at room temperature and used immediately whenever they are needed. Media that have not been autoclaved should be kept in the refrigerator and allowed several hours to warm up to room temperature before use, otherwise troublesome air bubbles will form during dissection of the oocytes and handling of their nuclei. Once a flask of medium has been opened and used the remaining contents should be disposed of and not reused. Bacterial contamination of lampbrush preparations is common, but it is quite unnecessary and can be avoided with a little care and good laboratory technique.

(iii) Isolation of the oocyte nucleus

Before commencing the operations detailed in this and the following section check that all the equipment is ready and to hand. Speed and smoothness of operation are the essence of success in working with lampbrush chromosomes, and this means being fully prepared for the job.

(1) Adjust the position of the light on the dissecting microscope such that it is focused to a thin bright beam shining at an angle of about 20° directly on to the centre of the microscope stage.
(2) Take a small piece of ovary and place it in 5 : 1 K/NaCl in an embryo cup on the stage of the dissecting microscope. Use a final magnification of about 10× at this point in the procedure.
(3) Take an observation chamber and place it on the stage of the dissecting microscope well out of the way at the back of the stage.
(4) Fill the observation chamber with dispersion medium such that there is a substantial convex (upwards) meniscus. The meniscus is important for two reasons. First, it acts as a lens and helps to concentrate the light on to the bottom of the chamber. Secondly, the preparation will eventually be covered with a coverglass and if there is not sufficient excess liquid in the chamber before the coverglass is added then air bubbles and turbulence in the chamber will result when the coverglass is dropped into place and the chromosomes will be severely damaged.
(5) Select an oocyte and measure it. With a pair of number 4 forceps grasp the oocyte near to where it is attached to the remainder of the ovary. The oocytes that are easiest to work with in nearly all species of amphibian are between 0.8 and 1.0 mm in diameter. Larger oocytes usually have rather contracted and featureless lampbrush chromosomes. Smaller oocytes are more difficult to handle, although their chromosomes are usually much

longer, more loopy, and generally more useful for lampbrush studies. Oocytes of less than 0.5 mm diameter are usually quite difficult to handle.

(6) Stab the oocyte through its follicle wall with the dissecting needle. Remember that the point of this needle should be polished to be smooth and round and just sharp enough to penetrate the follicle wall. The cytoplasm will immediately start to flow out and it can be encouraged to do so by applying gentle pressure with the needle on the side of the oocyte opposite to that which has been stabbed. Watch the flow of cytoplasm and at one point it will be interrupted by a clear patch that will mark the position of the oocyte nucleus. It is much easier to detect the nucleus at this stage than to find it at a later stage.

As soon as the nucleus is visible put down the needle and pick up the finely drawn Pasteur pipette. Hold the pipette with the bulb between the ball of the thumb and the base of the first finger and the stem between the tip of the second finger and the back of the tip of the third finger. Fill it with fluid, balancing the pressure so that it remains full right to the tip when no pressure is applied.

(7) Using very gentle pressure suck the nucleus up into the pipette and pump it in and out until it is completely free of yolky cytoplasm. When doing this be particularly careful about air bubbles. If the nucleus touches a bubble in the dish or near the mouth of the pipette, then it will burst and should be abandoned.

(8) When the nucleus is clean, pick it up with the pipette in such a way that it rests somewhere in the first centimetre of the pipette tip. The nearer the tip the better, but of course it must not come in contact with the air/water interface at the mouth of the pipette.

(9) Immediately move the embryo cup off the stage of the dissecting microscope and move the observation chamber into position. Focus down to the level of the observation chamber and then poke the mouth of the pipette beneath the surface of the liquid in the chamber. Transfer the nucleus into the chamber, watching the actual transfer with the dissecting microscope. The trickiest part of this procedure is holding the pipette with the nucleus in it while the embryo cup is replaced by the chamber on the microscope stage and the microscope refocused. The secret of success is the proper holding and balancing of pressure in the pipette. When putting the nucleus into the chamber try to transfer as little isolation medium as possible. Naturally, if a particular dispersion medium has been selected to encourage dispersal of the nucleoplasm then it will be counterproductive to dilute this medium by transferring over substantial amounts of isolation medium with the nucleus.

(10) When the nucleus is safely in the observation chamber, zoom up the microscope magnification to between $25\times$ and $30\times$ and focus very carefully just above the level of the top of the nucleus; pick up the no. 5 fine forceps and the tungsten needle.

(11) Press very gently on the top of the nucleus and to one side with the finest forceps, take a firm but not deep bite of the nuclear membrane, and immediately lift the nucleus up off the bottom of the chamber and into the level of focus of the microscope.

(12) Bring the point of the tungsten needle near to the point at which the nuclear membrane has been grasped by the forceps, prick the membrane, and tear it down round and away such that a large hole is made on the front side of the nucleus with respect to the operator. Continue moving the needle round the nucleus so as to lift it up under the shafts of the forceps with the hole pointing downwards. Then hold it there for a few seconds as the contents spill out onto the bottom of the chamber. If the operation has been successful the contents of the nucleus will spill out in the form of a single round ball of nucleoplasm and fall cleanly away from the nuclear membrane and the forceps. If the operation has not been successful then the contents of the nucleus will stick to everything and will streak out across the chamber as the forceps and needle are withdrawn.

(13) When the operation has been completed, place the chamber on the bench, take a scrupulously clean 22 mm or 18 mm square coverglass, hold it perfectly level about 1 cm above the hole in the observation chamber, and drop it squarely into position. It is essential not to lay the coverglass into position but to drop it directly from above.

(14) Lightly drop a filter paper over the top of the observation chamber and blot away excess liquid from around the top coverglass. Take care not to move the coverglass.

(15) Seal the edges of the coverglass using vaseline or paraffin grease and a hot wire. Preparations that have been made in chambers with slides forming their bases and are to be made permanent should not be sealed with vaseline. Instructions for dealing with preparations of this kind are to be found later in this Chapter (see Section III.E(ii), step (6)).

Preparations can be examined right away with a microscope, although it will probably not be possible to see much of the chromosomes until the nucleoplasm has dispersed. However, actually watching the process of dispersal gives a good impression of the arrangement of structures in the nucleus and it shows how ribonucleoprotein granules and other 'chromosome products' lie with respect to particular chromosome regions. Dispersal may take anything from 5 min to an hour or more. However, if the nucleoplasm is still clumped after about 30 min then it is probably worth considering some modifications to the dispersion medium: addition of more calcium, cytochalasin-B or formaldehyde, or reduction in ionic strength. There are no hard and fast rules about sap dispersal. The conditions that produce it are generally the same for all oocytes from a particular animal, but the rate at which it happens varies widely from one animal to another. Material from one newt may give preparations that disperse in less than a minute in 0.1 M 5 : 1 K/NaCl + 10^{-4} M CaCl$_2$, whereas those from another newt may take up to half an hour to disperse fully in the same medium. In an extreme case, nucleoplasm from a plethodontid salamander took over 12 h to disperse and the process could not be hurried by any modifications of the medium that were not harmful to the chromosomes. However, after that time the chromosomes were fully dispersed and still in good condition.

(iv)   Further hints on handling oocyte nuclei

The faster you can work the better. Expert lampbrushers usually take less than 1 min from selection of the oocyte to dropping the coverglass over the completed preparation. Practice makes perfect! Undergraduate students faced with the task and provided with favourable material and good equipment and media usually manage to make some good lampbrush preparations in the course of a half-day laboratory session. It takes a little longer to be consistently successful.

Some newts have particularly sticky nuclear membranes or naturally fluid nucleoplasm. Either of these characteristics can make the nuclei very hard to handle. The best advice in cases like this is pick another newt. Unfortunately, oocyte nuclei from *N. viridescens* and from *X. laevis* always show a tendency to stick to glass surfaces. The only way to overcome this problem is to keep the nucleus constantly on the move and work fast. Other tricks include siliconizing of all glassware except the observation chamber, and deliberately leaving a little cytoplasm on the nucleus before transferring it to the chamber. The latter step of course risks cytoplasmic contamination of the final preparation. To avoid this the nucleus should first be placed well to the side of the chamber and then picked up and lifted to the middle of the chamber for removal of its membrane so that the contents are spilled out on to a clean area on the base of the chamber.

Give an isolated nucleus 10–20 s to swell after it has been placed in the observation chamber and before attempting to remove its membrane. This allows a space to form between the chromosome mass and the nuclear membrane and enables the membrane to be grasped in the forceps without damage to the chromosomes. But don't wait too long. The nucleus swells rapidly and uncontrollably after isolation and it will soon burst. If the nucleus does burst or can be seen to have a leak through which nucleoplasm is escaping, then abandon it immediately.

Take pains to maintain a high standard of cleanliness at all stages of the procedure. This applies particularly to the medium in which the nuclei are removed from the oocytes and to the forceps and needle used for removing the nuclear membrane. Inevitably the isolation medium soon becomes messy with cytoplasm and yolk from punctured oocytes. Change it frequently, otherwise the final lampbrush preparations will have more yolk granules than chromosomes! Each time a preparation is made the points of the forceps and the tungsten needle will pick up some unwanted material, even if it is only nuclear membrane. Clean the forceps between preparations with a tissue. Clean the tip of the needle between preparations by holding it briefly in a hot bunsen flame. Never flame the points of the forceps as this will soften and generally ruin them.

Never make up media in bottles that have been used for biochemicals, culture media, or the like, or that have screw tops with cardboard or rubber inserts. Lampbrush chromosomes are exceedingly sensitive to contaminating biochemicals and to proteolytic enzymes, nucleases, and detergents in particular. So use only the cleanest glassware and cap it with foil, glass, fresh parafilm, or plastic.

Lampbrush chromosome preparations can be left for several days in the refrigerator provided that they are properly sealed and reasonably sterile. In one extreme case in the authors' laboratory some preparations were kept unfixed, in excellent condition, and without appreciable change in appearance for 14 months.

C. Looking at freshly isolated lampbrushes

Perhaps the hardest part for complete newcomers to this technique working alone is knowing when they have got things right. The important basic points to attend to in this regard are as follows:

(a) Have confidence in the dissection media. Know that they have the right pH and are free from contamination with biochemicals or detergents.

(b) Have confidence in the equipment. Are the chambers, forceps, needles, and any other items with which the oocytes, the nuclei, or their chromosomes come into contact, clean and uncontaminated?

(c) Select oocytes of the right size range: 0.8–1 mm diameter.

Given these conditions, properly isolated lampbrush chromosomes should show the following features:

(a) They should disperse to lie flat on the bottom of the chamber within less than half an hour.

(b) They should have large numbers of lateral loops projecting from the axes of the chromosomes all along their lengths.

(c) The loops should be clearly distinguishable and individually resolvable at magnifications of 25× and above.

(d) The chromosomes and their loops should show vigorous Brownian movement when examined at magnifications of 40× or above.

If the chromosomes have very clearly defined axes and their loops are short fuzzy structures emerging from substantial chromomeres, then the preparation has probably been made from too large an oocyte. In larger oocytes of most species the loops have partially withdrawn into the chromosome axis and the axis itself has contracted and thickened. Figures 6.3 and 6.4 show two chromosomes both from the same animal. The chromosome in Figure 6.3 is from a 1.5 mm oocyte in which substantial chromosome contraction has taken place. That in Figure 6.4 is from a 0.8 mm oocyte where the lampbrush loops and the chromosomes are still maximally extended. Note that both figures are reproduced at the same magnification.

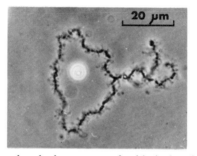

Figure 6.3   A lampbrush chromosome freshly isolated and unfixed from an oocyte of 1.5 mm diameter (*Triturus cristatus carnifex*) in which substantial chromosome contraction has taken place and the lampbrush loops have largely withdrawn into the chromosomes' axes. Phase contrast photomicrograph

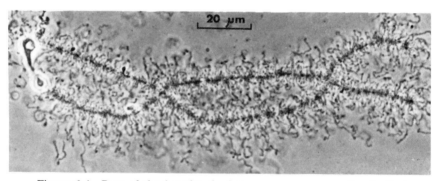

Figure 6.4   Part of the lampbrush chromosome corresponding to that shown in Figure 6.3, freshly isolated and unfixed, photographed and reproduced at the same magnification as in Figure 6.3, but from an oocyte of 0.8 mm diameter in which the chromosomes and their loops are still maximally extended and active. Phase contrast photomicrograph

If the chromosomes appear as loopless threads, then the dispersion or isolation media are most likely contaminated with something that strips the ribonucleoprotein matrix from the loops. If they appear stiff and highly refractile with no evident Brownian motion on the loops, and they have not dispersed properly, then the dispersion medium is most likely to be too acid or has too much calcium. In general and unavoidably subjective terms, a freshly isolated lampbrush chromosome from a 0.8–1 mm diameter oocyte of a crested newt is an aesthetically pleasing object, and it is very plain to see if something has gone wrong.

Figures 6.5–6.9 show parts of four preparations each from the same animal, all photographed at the same magnification and under precisely the same conditions. Figures 6.5 and 6.7 show chromosomes that look as they should when everything has gone just right. Figure 6.6 shows the same chromosome as in Figure 6.5 after standing for a few minutes in an isolation medium that was too acid (slightly below pH 5). Figure 6.8 shows the same chromosome as in Figure 6.7 after standing for about 5 min in a medium that was contaminated with some component of calf serum, having been made in a serum bottle with a rubber-sealed cap that had supposedly been thoroughly washed beforehand. The offending substance in this case was almost certainly ribonuclease. Figure 6.9 shows part of a preparation that was made from a nucleus that was already leaking when the author started to remove the nuclear membrane. The entangled mass of chromosomes and many bits of broken loops are quite characteristic of this kind of mishap.

D. Identification and mapping of chromosomes

The definitive work on lampbrush chromosome identification and mapping is Callan and Lloyd's (1960) paper on the lampbrush chromosomes of crested newts. This paper provides a clear photographic and diagrammatic record of all the many and varied features of newt lampbrush chromosomes. It gives a good account of how to locate and recognize centromeres, how to measure chromosome arms, how to record the positions of site-specific objects, and how to interpret the shapes of the chromosomes and some of the extraordinary patterns of fusion and association that form quite consistently at certain places in the chromosome set. An investigator coming new to lampbrushology and intent upon familiarizing him/herself with the lampbrush chromosomes of any organism should first study Callan and Lloyd's paper thoroughly and with minute attention to detail. In a later publication Callan and Lloyd (1975) summarize and give references for working maps of the lampbrush chromosomes of all species of newt for which the chromosomes had been described up to that date.

When exploring the lampbrushes of any species it helps to remember that these chromosomes are in diplotene with half bivalents joined to one another

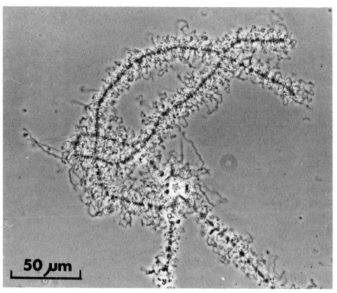

Figure 6.5   Part of a freshly isolated and unfixed lampbrush chromosome photographed in phase contrast immediately after isolation and dispersal

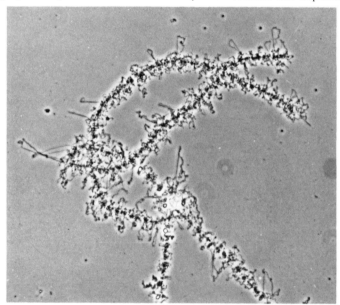

Figure 6.6   The same lampbrush chromosome as shown in Figure 6.5, 10 min after
isolation. The isolation medium was slightly too acid (just below pH 5), and this has
had the effect of causing the loops and the chromosomes' axes to contract slightly
and stiffen, giving the chromosome a highly refractile and fixed appearance that is
quite unlike the loose and delicate appearance of a freshly isolated lampbrush lying
in a properly balanced medium

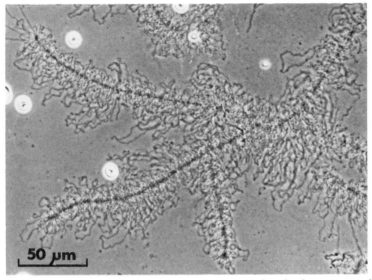

Figure 6.7 Phase contrast picture of a part of a lampbrush chromosome freshly isolated and unfixed. The chromosomes were isolated in a medium that had been stored in a supposedly clean and sterile bottle that had once contained calf serum. The cap of the bottle was lined with the original rubber seal. The chromosomes in the picture have the appearance of normal well isolated lampbrushes. They were photographed just 2 min after removal of the nuclear envelope

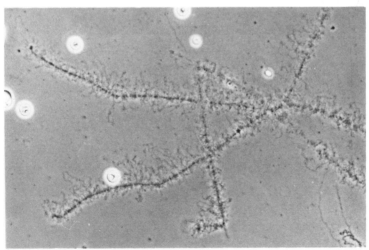

Figure 6.8 Phase contrast picture of the same chromosomes as in Figure 6.7, 5 min after isolation. The chromosomes and their loops now have a stripped appearance, much of the loop RNP having been removed by a contaminant in the isolation medium. The contaminant is presumed to have been ribonuclease left associated with the serum bottle and its cap in which the isolation medium was stored prior to use. This kind of effect is typical of the effects of either ribonuclease or certain proteolytic enzymes

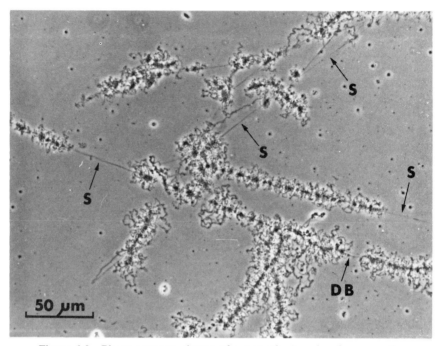

Figure 6.9   Phase contrast picture of a part of a lampbrush preparation
that was made from a nucleus that was already leaking when the operator
started to remove the nuclear envelope. The chromosomes are extensively
broken and tangled, and many stretched loops (S) and double-bridge
breaks (DB) (Callan and Lloyd, 1960) are evident

at chiasmata, and it is also most useful to know the chromosome number
and the metaphase karyotype of the organism in question. The interpretation
and mapping of a set of lampbrush chromosomes in a freshly made preparation involves careful and painstaking microscopy with constant switching
between high and low power objectives. The greatest problem for the beginner stems from the fact that the bivalents rarely come to lie completely
separate from one another, no matter how good the preparation. There are
bound to be overlaps, and it is sometimes hard to tell what is an overlap
between two different bivalents, what is a chiasma between half bivalents,
and what is a fusion. Fusions can be particularly troublesome since they can
involve non-homologous parts of two half bivalents – a situation that cannot
at once be distinguished from a chiasma – and non-homologous parts of
completely different bivalents. Callan and Lloyd's (1960) paper gives some
excellent examples of complex fusions and should be helpful to persons
confronted with these kinds of situations for the first time.
    As a rule it is best to examine a preparation with objective magnifications

of between 10× and 20× to determine the general lie of the chromosomes with respect to one another, and then switch from time to time to 40× to study focal planes and so distinguish overlaps from true associations.

For the purposes of recording chromosome lengths and the positions of distinctive marker structures it is simplest and most effective to use a *camera lucida* drawing attachment on an inverted microscope, ideally in conjunction with a computerized digitizer or graphics tablet. Lampbrush chromosomes are very long, thin, low contrast objects, and they are much more easily drawn than photographed under low magnifications.

E. Permanent preparations

When lampbrush preparations are dispersed over a flat glass surface in a medium that preserves their life-like form, the slightest disturbance of the surrounding medium will move the chromosomes, stretch them, and produce multiple breaks in loops and chromosome axes. Therefore the major problem that had to be overcome in the quest for good permanent preparations of lampbrushes was that of sticking the chromosomes and their loops securely to the underlying surface so that they would not be at all damaged or disturbed as the slides were passed from one stage to another of a histochemical procedure. In the early days this was accomplished by allowing the chromosomes to disperse completely and lie flat on the base of an observation chamber and then exposing them to vapour from either acidified formalin, acetic acid or ethanol. This had the effect of fixing and solidifying the thin layer of nucleoplasm surrounding the chromosomes and they became sufficiently trapped to withstand the turbulence involved in fixing, staining, dehydrating, and mounting. A version of this simple method will be described later. However, it is by no means dependable and losses are common.

A much better method and one that guarantees complete success was introduced as a modification of the 'Miller spreading technique' for preparing chromosomes and chromatin for visualization of transcription (see Chapter 7). The essential step is the centrifugation of the chromosomes on to the bottom of the observation chamber such that they and their loops come to lie completely flat and securely stuck to the glass surface. After this, no amount of disturbance of the surrounding medium will damage or move them, although at this stage they can still be severely damaged if allowed to pass through an air/water interface.

Permanent fixed preparations of lampbrush chromosomes are likely to be needed for two main purposes. In the first place an investigator may wish to fix and stain some preparations from a particular animal and keep them as part of a research record or teaching collection. In the second place an investigator may wish to fix and make permanent some preparations that will subsequently be exposed to certain experimental treatments. A common example of a need of this kind is in experiments involving *in situ* hybridization of radioactively labelled nucleic acids to lampbrush chromosomes and subsequent autoradiography. The preparation must be fixed, dried, processed through a number of experimental steps, then dried again, and finally coated with a nuclear track emulsion. What follows is a simple but rather risky method for sticking chromosomes to the base of an observation chamber and fixing them histologically, and then the more complex and dependable centrifugation method.

It should be made clear at this point that the method used to stick a lampbrush chromosome down on a slide does not limit what can be done with it thereafter. However, preparations that have been stuck down by centrifugation are likely to be much more amenable to good microscopical analysis simply because their loops will all be laid down perfectly flat on either side of the chromosome axis and everything will be in the same focal plane.

Two important basic rules apply to protocols for making permanent preparations of lampbrush chromosomes. First, avoid formaldehyde if the preparation is to be used for any subsequent biochemical work such as treatment with enzymes or labelled antibodies. The instructions given here offer the alternative of using either formaldehyde or ethanol as the primary fixative. Use formaldehyde only for straightforward cytology or cytochemistry. Use ethanol for everything else. Secondly, if a preparation is to be stained and mounted directly then it should not be air dried. If it is to be processed for biochemical work, *in situ* hybridization, or autoradiography then it should be air dried immediately after it has been fixed. Air drying ensures that the chromosomes are well and truly stuck to the slide, and this is most important considering the rigours and complexity of, for example, the *in situ* hybridization technique. Preparations should definitely be air dried before dipping in nuclear track emulsion.

(i)   Sticking and fixing without centrifugation

(1) Make a lampbrush chromosome preparation in the normal way in an observation chamber constructed with a microscope slide as its base and in a medium that has previously been tested and is known to allow good dispersal of the chromosomes in a reasonable time. Either bored slide or plastic card chambers can be used. Do not cover the preparation with a coverglass, but just stand it in a moist chamber at room temperature, completely undisturbed, for as long as is known to be needed for good dispersal.

(2) When dispersal is judged to be complete, and not before, pour on to the filter paper in the moist chamber a little of a solution containing either 1% acetic acid and 20% formaldehyde, or 70% ethanol. Put the lid back on the moist chamber and wait for 10–15 min, during which time the vapour

from the fixative will diffuse into the isolation medium and the chromosomes will become fixed and attached to the base of the observation chamber.

(3) Cover the preparation with a 22 mm square coverglass.

(4) Remove the observation chamber from the moist chamber.

(5) Place a filter paper over the observation chamber and carefully blot away excess medium around the coverglass.

(6) Press down gently on the coverglass to make sure that it is firmly in position.

(7) Place the observation chamber in a Petri dish.

(8) Carefully pour into the Petri dish a solution of 5% formaldehyde or 50% ethanol until the entire observation chamber is completely immersed.

(9) Using a dissecting needle or the tips of a pair of forceps slowly push the coverglass aside, allowing the fixative to mix with the isolation medium in the chamber.

(10) Wait about 1 min.

(11) Gently move the slide about in the dish to obtain good mixing of the formaldehyde or ethanol and the contents of the observation chamber.

(12) Leave for 1 h.

(13) Push the coverglass back into position over the observation chamber and press it down firmly.

(14) Pour off the fixative and replace it with 70% ethanol, making sure the coverglass does not move.

(15) When the slide is fully immersed in 70% ethanol gently push the coverglass aside, wait 1 min, swish the slide about to encourage mixing, and then leave for 10 min.

(16) Replace the coverglass.

(17) Remove the observation chamber from the dish.

(18) Hold it in a 500 ml beaker of 70% ethanol.

(19) Allow the coverglass to fall off in the ethanol.

(20) Keeping the slide immersed at all times, use a scalpel or razor blade to prise the two slides of the observation chamber apart or to remove the plastic card, as the case may be. Remember that at this stage it is important to know which side of the slide carries the preparation and mark it with a diamond pencil if this has not already been done.

(21) If the chromosomes have not already been lost up to this point then they will now be well and truly stuck to the slide and it is possible to proceed with reasonable care and confidence in the normal way. The slide can be passed through other solutions, stains, etc., like any other histological preparation, or it can be passed through two changes of 95% ethanol and air dried in preparation for *in situ* hybridization or autoradiography.

(ii)   *Sticking and fixing with centrifugation*

The centrifugation technique requires the construction of special centrifuge baskets that will accept the normal or double-slide observation chambers. It also requires a centrifuge that can be started in a slow and controlled manner and that is capable of accelerating to at least 3000 rev/min when loaded with two or more centrifuge baskets. A diagram of a basket constructed for use with a Sorvall GLC-1 centrifuge fitted with an HL-4 rotor is shown in Figures 6.10 and 6.11. The basket is most easily constructed from good quality brass. The same design, appropriately modified to take account of the width and dimensions of the pivot slots on the rotor and the space available for swing-

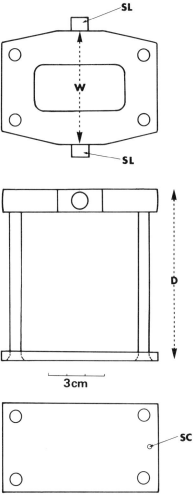

Figure 6.10

Figures 6.10 and 6.11   Scale drawing (reproduced at half size) and photograph of a centrifugation basket suitable for spinning lampbrush observation chambers on the HL-4 rotor of the Sorvall GLC-1 centrifuge. The basic design can be modified to suit almost any swing-out rotor. The three parameters that are likely to need changing for other makes of centrifuge are the sizes of the suspension lugs (SL), the width between the lugs (W), and the depth of the basket (D). A small screw (SC) may be inserted into one end of the platform to help locate the observation chamber and hold it in position while the centrifuge is accellerating

ing, can be used for making baskets for several different centrifuges. Baskets of this kind have been made for the following centrifuges: Sorvall GLC-1 (HL-4 rotor), Sorvall RC5-B (HS-4 rotor), MSE Super Minor or Chillspin, and International PR-2 (269 rotor).

The Sorvall RT-6000 bench-top refrigerated centrifuge is best of all for spinning lampbrush preparations. It requires no special baskets and takes up to 16 preparations at a time.

The procedure for centrifuging lampbrush preparations is as follows:

(1) Make a lampbrush preparation in the normal way in an observation chamber with a microscope slide as its base and in a medium that has previously been tested and is known to allow good dispersal of the chromosomes in reasonable time. Either bored slide or plastic card chambers can be used.

(2) Cover the preparation with a coverglass as soon as it has been made.

(3) Place the preparation in a moist chamber which is standing on crushed ice, and leave it completely undisturbed for as long as is known to be needed for good dispersal.

(4) When dispersal is judged to be complete, remove the preparation from the moist chamber and place it on the bench.

(5) Carefully blot away all excess fluid from around the coverglass.

(6) Fix the top coverglass in position and seal around its edges either by applying vaseline or paraffin grease with a hot wire or by sticking a wide piece of Sellotape (Scotch tape) over the entire coverglass. Take great care not to move the coverglass while carrying out these sealing operations.

(7) Place the observation chamber on the platform of one of the centrifuge baskets and put the basket in position on the centrifuge rotor. Counterbalance the rotor by placing another preparation or an empty observation chamber on the opposite basket.

(8) Start the centrifuge as slowly as possible, gradually working up to 3000–4000 rev/min (1500–2000g) over a period of about 3 min. It is most important to avoid a rapid surge of acceleration at the start of the spin as this may sweep the chromosomes over to one side of the observation chamber. Centrifuge at 1500–2000g for 15 min.

(9) Remove the chamber from the centrifuge. Immerse it in 0.1 M 5 : 1 K/ NaCl and use a razor blade to prise apart the two slides of the observation chamber or to remove the plastic card together with the top coverglass. Keep the chamber all the while completely immersed in 5 : 1. Gently swish the slide bearing the preparation around in the 5 : 1. This step helps to wash away the nucleoplasm that surrounds the chromosomes.

(10) Immediately transfer the preparation to whatever fixative is appropriate in relation to subsequent treatment. This is likely to be either 70% ethanol or 5% formaldehyde buffered to pH 7.

(11) Remove the slide, mark it with a diamond pencil on the side that carries the preparation, and transfer it immediately to a Coplin jar full of fresh fixative.

Subsequent treatment will depend on what is to be done with the chromosomes in the long run. If the preparation is to be stained and mounted directly, then the protocol given in Section V of this chapter should be followed. If the preparation is to be used for any biochemical approach, *in situ* hybridization, or autoradiography, then it is advisable to fix it first in 70% ethanol for between 15 min and 1 h, then pass it through two 10 min changes of 95% ethanol and air dry it. Preparations can be stored in the refrigerator until such time as they are needed for the next step in the experimental procedure. For the best results they should be stored in sealed boxes with a vial of desiccant.

IV. Modifications of the basic technique for special purposes

A. The actions of enzymes on lampbrush chromosomes

Over the years a number of simple but highly significant experiments have been carried out in which lampbrushes have been dissected directly into media containing proteases and nucleases (Callan and Macgregor, 1958; Macgregor and Callan, 1962; Gall, 1963; Gould *et al.*, 1976). This is a useful experimental approach, one that can be carried out with the minimum of equipment and fuss, and one that is particularly appealing to new students and therefore suitable as a teaching exercise in the sense that it yields instant results in a spectacular manner.

The essence of the approach is the basic lampbrush technique with nothing changed except for the final dispersion medium that goes into the observation chamber. Chromosomes are merely dissected into a special medium containing a specific enzyme, and then the effects of the enzyme are watched and photographed with a phase contrast microscope.

A suitable dispersion medium for studying the actions of enzymes on lampbrush chromosomes of *T. cristatus* or *N. viridescens* is as follows: 0.1 M 5 : 1 K/NaCl; 0.01 M Tris buffer, pH 7–7.5; 0.001 M $MgCl_2$; 0.05 mM $CaCl_2$; 0.001 M mercaptoethanol. The mercaptoethanol must be made up and added fresh each day. Place the medium in a vial with a loosely fitting cap and degas it under vacuum for about 15 min (a vacuum pump of the kind that fits to a mains water tap is adequate for this purpose). Then fill the vial with nitrogen and cap it tightly. In this medium, Tris buffer holds the pH at a little over 7, which is appropriate for lampbrush chromosomes and for a useful range of enzymes. Magnesium ions are essential as a cofactor for many nucleases. Mercaptoethanol acts as a reducing agent and prevents oxidation of chromosomal proteins; however, it also breaks nucleic acids in the presence of oxygen, hence the degassing and nitrogen replacement procedure.

Finally, in this context it should be said that the actions of common proteases such as trypsin and pronase and of ribonuclease and deoxyribonuclease can easily be studied without the use of special media. Any of these enzymes will have their effect if simply added at about 0.1 mg/ml or less to a normal dispersion medium, although DNase may require the addition of Mg^{2+} to be fully effective.

With regard to restriction endonucleases some experimentation around the basic medium suggested above is likely to be needed and in this regard account will have to be taken of the optimal conditions for the action of each particular endonuclease.

B.   Radioisotope incorporation into germinal vesicles

Amphibian oocytes, and doubtless those of many other kind of animal as well, can be incubated *in vitro* in the presence of radioactively labelled nucleotides or other precursors, and the incorporation of these substances into the nucleic acids and proteins of lampbrush chromosomes can be studied with normal lampbrush techniques followed by autoradiography. Once again the procedures are remarkably simple and straightforward. The only special information needed here relates to the media recommended for *in vitro* incubations.

A medium that has been found to be suitable for short term incubations of oocytes lasting not more than 4 h is full strength Steinberg's solution which consists of the following: 0.05 M NaCl; 6.7 × 10^{-4} M KCl; 3.4 × 10^{-4} M Ca(NO$_3$)$_2$.4H$_2$O; 8.1 × 10^{-4} M MgSO$_4$.7H$_2$O; 4.6 × 10^{-5} M Tris; 2.5 × 10^{-3} M HCl.

For longer incubations, lasting up to 1 week, with daily changes of medium and near sterile conditions, amphibian minimal essential medium with added serum gives good results and is made up as follows:

(1) Use one packet of Gibco MEM (cat. no. 072-1100), recommended for making up to 1 l.

(2) Dissolve this in 400 ml of glass-distilled water.

(3) Dissolve 5 g of lactalbumin hydrolysate (Gibco cat. no. 152-0025) in 500 ml of glass-distilled water.

(4) Mix MEM and lactalbumin hydrolysate solutions together.

(5) Add 100 ml of fetal calf serum (Gibco cat. no. 629H1).

(6) Add 12.5 ml of stock (100×) antibiotic/antimycotic (Gibco cat. no. 043-5240). This mixture contains penicillin,-10^4 units/ml, streptomycin,-10^4 mg/ml; Fungizone, 25 mg/ml.

(7) Add 12.5 ml of kanamycin (100 µg/ml; Gibco cat. no. 043-5160).

(8) Add 2 g of sodium bicarbonate.

(9) Make up to 1250 ml.

(10) Adjust to pH 7 with dilute HCl, bearing in mind that the pH will rise a little after filter sterilization.

(11) Pass through a glass wool prefilter.

(12) Filter through a Millipore 0.45 µm pore size bacterial filter into sterile glassware and store frozen in convenient sized aliquots.

The antibiotic/antimycotic components may be omitted for short term culture provided that the medium is carefully filter sterilized and changed from time to time during the incubation period.

Another successful defined nutrient medium for short term *in vitro* maintenance of *Xenopus laevis* oocytes has also been worked out by Wallace *et al.* (1973) (see also Eppig and Dumont, 1975) and is entirely suitable for use with material from other amphibian species. It is generally referred to as 'OR2' and is made up as follows:

(a) Stock solution A: NaCl, 48.21 g; KCl, 1.86 g; Na_2HPO_4, 1.42 g; HEPES (Sigma cat. no. H3375), 11.92 g; 1 M NaOH, 33 ml; made up to 1l with distilled water.

(b) Stock solution B: NaCl, 2.65 g; $MgCl_2.6H_2O$, 2.03 g; made up to 1l with distilled water.

(c) Working solution: stock A, 100 ml; distilled water, 300 ml; phenol red, 2 ml; stock B, 100 ml; polyvinylpyrrolidone (Sigma cat. no. PVP-40) stock solution (12.5 g) in 250 ml water, filter sterilized and stored frozen, 10 ml; antibiotic/antimycotic (Gibco cat. no. 043-5240), 10 ml; made up to 1 l with distilled water.

The simplest and best procedure for incubations in any of these media with the object of obtaining incorporation of radioisotopically labelled compounds into oocytes and lampbrush chromosomes is to evaporate a small amount of the labelled compound, usually 20–100 μCi, in the bottom of a sterile embryo cup and then mix it with medium to give a final concentration of the isotope of about 100 μCi/ml, using sterile technique throughout. A piece of ovary is then removed from the animal, washed through several changes of sterile medium without the labelled compound, and placed in the dish of radioactive medium. If the incubation is to last for more than 4 h then at least 1 ml of the labelled medium should be used for a bunch of about 50 oocytes, the medium should be kept on the move using a gentle shaker. For longer incubations the medium should be changed at 12 h intervals or less.

After incubation the oocytes should be washed in cold (i.e. non-radioactive) medium and the lampbrush preparations should be made in the dispersion medium recommended for use in enzyme digestion experiments (see Section IV.A). The preparations should be fixed in formaldehyde or ethanol, centrifuged, and then washed for 5 min at 5°C in a freshly prepared 5% solution of trichloroacetic acid. After that they should be washed, dehydrated, and dried in the normal way for autoradiography.

It is worth mention that medium OR2 can be used with collagenase to dissociate amphibian ovaries into individual oocytes that are completely stripped of their surrounding follicle epithelium. This approach is of considerable value where an investigator wishes to carry out biochemical studies on oocytes of a particular size without contaminating somatic cells. The same technique can be applied for dissociating other amphibian tissues. Tissues are first rinsed in OR2 and then incubated at 25°C in OR2 + 0.2% collagenase (Sigma cat. no. CO130; 230 units/mg) with vigorous agitation. *Xenopus* ovary usually disintegrates into a single cell suspension in less than 1 h under these conditions.

V.   Staining lampbrush chromosomes

There are three useful and dependable ways of staining lampbrush chromosomes. The stains involved are Giemsa, Coomassie Blue, and Heidenhain's haematoxylin. Giemsa and Coomassie Blue are normally used and are to be recommended for staining preparations that have at some stage been air dried. For example, one or other of these stains is routinely used for staining chromosomes in autoradiographs after development and fixing (see Chapter 9, Section XI). Giemsa and Coomassie Blue can also be used for staining lampbrush chromosomes immediately after fixation, but they are not the best stains for this purpose. Haematoxylin gives far superior results but only with preparations that have never been air dried; it is of no use whatever for staining chromosomes in autoradiographs. The rule, then, is to use haematoxylin if the staining is to be done immediately after fixation and all that is wanted is handsomely stained well preserved permanent lampbrush preparations. For all other purposes use Giemsa or Coomassie Blue. Only the haematoxylin method will be given here. Giemsa and Coomassie Blue staining methods are given in Chapter 9, Section XI.

(1) Make lampbrush preparations in the normal way using either slide-bottomed or plastic card observation chambers.

(2) Centrifuge the preparations.

(3) Separate the slides or remove the plastic card discs under 0.1 м 5 : 1 K/NaCl.

(4) Swish the slides around gently in the 5 : 1 for a few moments to wash away nucleoplasm surrounding the chromosomes.

(5) Transfer the slides, taking great care to keep the area over the preparation wet, to a solution of 5% formaldehyde buffered to pH 7, and leave them in this for 1 h.

(6) Wash the slides in running tap water for 30 minutes.

(7) Pass the slides through 50%, 70%, 95%, and 100% ethanol, leaving them 10 min in each.

(8) Place the slides in a staining rack in a dish of xylene with a magnetic stirring bar. This step is needed to remove all traces of wax that may have been used in making double-slide observation chambers.

(9) Take the slides back through the alcohols to distilled water.

Note that steps (7), (8), and (9) can be omitted if the preparations were made in plastic card chambers.

(10) Mordant in 3% iron alum for 12 h or overnight.

(11) Wash thoroughly in distilled water.

(12) Place in Heidenhain's haematoxylin for 6 h.

(13) Wash thoroughly in distilled water.

(14) Place in 0.5% iron alum for 1–2 min. The optimal time at this stage is best determined by trial and error, but in any case it will be short.

(15) Wash thoroughly in distilled water.

(16) Pass through the alcohol series, 50–100%, leaving the preparations 10 min in each.

(17) Pass through two changes of xylene and mount in Canada balsam.

Remember, the secret of success in this procedure is to make sure that the chromosomes are never allowed to dry. Keep the slides thoroughly wet at each transfer.

VI. Lampbrush chromosomes from animals other than newts

Whatever species is being investigated the would-be lampbrushologist will encounter two main problems. First, the germinal vesicle must be removed from the oocyte, cleaned, and transferred to an observation chamber. Secondly, the nucleoplasm must be made to disperse so that the chromosomes spread out and lie flat on the bottom of the chamber. The latter problem can usually be solved by experimenting with different dispersion media: lowering ionic strength, though not so much as to dissolve the ribonucleoprotein from the lampbrush loops; raising pH, though not above 7.5; introducing calcium, though not so much as to cause precipitation of the nucleoplasm or stiffening of the chromosomes; adding formaldehyde, but never in excess of 1% and never if preparations are to be used for biochemical experiments; or adding cytochalasin-B. If none of these measures, alone or in combination, are effective then the experimenter must either use initiative or give up!

The problem of finding, cleaning, and transferring the germinal vesicle can be a harder one, but in the authors' experience it can usually be done by some means or other. Amphibians present no real difficulty. Cytoplasm usually flows out through a tear in the oocyte, though a little persuasion may be needed in the form of a gentle squeeze. The germinal vesicles of amphibian oocytes are usually quite large and can be seen as they pass through the tear. Even if they cannot be spotted at this stage, then a little gentle pipetting of the cytoplasm will soon expose them and once located they can be picked up and cleaned by gently pumping them in and out of the pipette. After cleaning they usually remain clearly visible, mainly because they contain many hundreds of large extra nucleoli and these give the nucleus a bright glistening appearance. As a rule, careful watching, critical lighting, and skilful pipetting are enough to ensure location and successful isolation of an amphibian germinal vesicle.

The location and isolation of germinal vesicles from oocytes of other animals is usually much harder, but the same general approaches and rules apply. With reptiles, for example, the follicle wall is exceedingly thick (squamates and amphisbaneans only) and quite hard to tear. The cytoplasm is stiff and must usually be pushed out of the torn follicle by squeezing the back of the oocyte with the dissecting needle. The germinal vesicles of reptiles are always much smaller than those of amphibians and in squamates and most amphisbaneans there are few or no extra nucleoli, so that the nucleus when cleaned is a small highly transparent ball and is hard to see under any conditions. The trick with material like this is to locate the nucleus as a clear patch amongst the thick yolky cytoplasm and then literally pick the cytoplasm away from around it using the finest forceps and a tungsten needle but taking the greatest care not to touch the nucleus itself. Once clear of cytoplasm the only remaining problem is to make the transfer to the observation chamber without losing sight of the nucleus. Removal of the nuclear membrane presents no greater difficulty than with amphibian material so long as one can see what one is doing. Examples of studies of germinal vesicles and lampbrush chromosomes from a frog and a reptile can be found in Macgregor and Kezer (1970) (on the Pacific tailed frog, *Ascaphus truei*) and Macgregor and Klosterman (1978) (on the amphisbanean, *Bipes biporus*).

References

Barsacchi, G., Bussoti, L., and Mancino, G. (1970). The maps of the lampbrush chromosomes of *Triturus* (Amphibia Urodela). IV. *Triturus vulgaris meridionalis. Chromosoma (Berl.)*, **31**, 255–270.

Callan, H. G. (1966). Chromosomes and nucleoli of the axolotl *Ambystoma mexicanum. J. Cell Sci.*, **1**, 85–105.

Callan, H. G., (1982). Lampbrush chromosomes. *Proc. Roy. Soc. Lond. B*, **214**, 417–448.

Callan, H. G. and Lloyd, L. (1960). Lampbrush chromosomes of crested newts *Triturus cristatus* (Laurenti), *Phil. Trans. Roy. Soc. Lond. B*, **243**, 135–219.

Callan, H. G., and Lloyd, L. (1975). Working maps of the lampbrush chromosomes of Amphibia. In *Handbook of Genetics*, Vol. 4 (R. C. King, ed), Plenum, New York, pp. 57–77.

Callan, H. G., and Macgregor, H. C. (1958). Action of deoxyribonuclease on lampbrush chromosomes. *Nature (Lond.)*, **181**, 1479–1480.

Duryee, W. R. (1937). Isolation of nuclei and non-mitotic chromosome pairs from frog eggs. *Arch. Exp. Zellforsch.*, **19**, 171–176.

Eppig, J. J., and Dumont, J. (1976). A defined nutrient medium for the *in vitro* maintenance of *Xenopus laevis* oocytes. *In Vitro*, **12**, 418–427.

Flemming, W. (1882). *Zellsubstanz, Kern, und Zelltheilung*, F. C. W. Vogel, Leipzig.

Gall, J. G. (1954). Lampbrush chromosomes from oocyte nuclei of the newt. *J. Morph.*, **94**, 283–351.

Gall, J. G. (1963). Kinetics of deoxyribonuclease action on chromosomes. *Nature (Lond.)*, **198**, 36–38.

Gall, J. G., and Callan, H. G. (1962). ³H uridine incorporation in lampbrush chromosomes. *Proc. Natl Acad. Sci.*, **48**, 562–570.

Gould, D. C., Callan, H. G., and Thomas, C. A. (1976). The actions of restriction endonucleases on lampbrush chromosomes. *J. Cell Sci.*, **21**, 303–313.

Jamrich, M., and Warrior, R. (1982). Study of transcription and protein distribution on *Xenopus* lampbrush chromosomes. *J. Cell Biochem.*, Suppl. 6, 317 (abstract).

Kezer, J. (1982). Personal communication.

Lacroix, J.-C. (1968). Etude descriptive des chromosomes en ecouvillon dans le genre *Pleurodeles* (Amphibien, Urodele). *Ann. Embryol. Morphog.*, **1**, 179–202.

Macgregor, H. C. (1977). Lampbrush chromosomes. In *Chromatin and Chromosome Structure* (R. A. Eckhardt and Hsueh-jei Li, eds), Academic Press, New York and London, pp. 339–357.

Macgregor, H. C. (1980). Recent developments in the study of lampbrush chromosomes. *Heredity*, **44**, 3–35.

Macgregor, H. C., and Callan, H. G. (1962). The actions of enzymes on lampbrush chromosomes. *Quart. J. Microsc. Sci.*, **103**, 173–203.

Macgregor, H. C., and Kezer, J. (1970). Gene amplification in oocytes with 8 germinal vesicles from the tailed frog *Ascaphus truei* Stejneger. *Chromosoma (Berl.)*, **29**, 189–206.

Macgregor, H. C., and Klosterman, L. L. (1978). Observations on the cytology of *Bipes* (Amphisbanea) with special reference to its lampbrush chromosomes. *Chromosoma (Berl.)*, **72**, 67–87.

Mancino, G., and Barsacchi, G. (1966). Le mappe dei cromosomi lampbrush di *Triturus* (Anfibi Urodeli). II. *Triturus helveticus helveticus*. *Riv. Biol. (Perugia)*, **59**, 311–351.

Mancino, G., and Barsacchi, G. (1969). The maps of the lampbrush chromosomes of *Triturus* (Amphibia Urodeli). III. *Triturus italicus*. *Ann. Embryol. Morphog.*, **2**, 355–377.

Mancino, G., Barsacchi, G., and Nardi, I. (1969). The lampbrush chromosomes of *Salamandra salamandra* (L.) (Amphibia Urodela). *Chromosoma (Berl.)*, **26**, 365–387.

Mancino, G., Nardi, I., and Ragghianti, M. (1972). Structural correspondence between nucleolus and sphere organizing regions of the lampbrush chromosomes and secondary constrictions of the mitotic chromosomes. *Experientia (Basel)*, **28**, 586–588.

Nardi, I., Ragghianti, M., and Mancino, G. (1972). Characterization of the lampbrush chromosomes of the marbled newt *Triturus marmoratus* (Latreille, 1800). *Chromosoma (Berl.)*, **37**, 1–22.

Ruckert, J. (1892). Zur Entwicklungsgeschichte des Ovarialeies bei Selachiern. *Anat. Anz.*, **7**, 107–158.

Sommerville, J. (1977). Gene activity in the lampbrush chromosomes of amphibian oocytes. In *International Review of Biochemistry, Biochemistry of Cell Differentiation II* (J. Paul, ed), vol. 15. University Park Press, Baltimore, Md, pp. 79–154.

Vlad, M., and Macgregor, H. C. (1975). Chromomere number and its genetic significance in lampbrush chromosomes. *Chromosoma (Berl.)*, **50**, 327–347.

Wallace, R. A., Jared, D. W., Dumont, J. N., and Sega, M. W. (1973). Protein incorporation by isolated amphibian oocytes. III. Optimum incubation conditions. *J. Exp. Zool.*, **184**, 321–334.

7

Visualization of transcription
Contributed by
Aimee H. Bakken and R. S. Hill

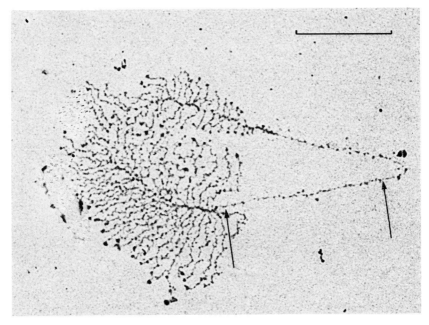

This electron micrograph shows a completely normal *Xenopus* ribosomal gene transcription unit and its associated non-transcribed spacer region. The transcription unit
is initiating and terminating at the proper locations, even though the frog ribosomal
DNA has been ligated to a bacterial plasmid at or close to the termination end of the
ribosomal gene. One entire *Xenopus* ribosomal DNA repeat unit was inserted into
a plasmid vector, pBR322 (shown between the arrows), and then cloned in bacteria.
The recombinant DNA was harvested and micro-injected into a *Xenopus* oocyte.
Note that the plasmid DNA assumes a nucleosomal configuration in the oocyte, using
the host cell's histones. Bar = 0.5 μm

The technique described in this chapter was introduced by Oscar Miller and his associates for studying the morphology and arrangement of transcriptionally active chromatin and its associated ribonucleoprotein (RNP) fibrils in the giant nuclei of amphibian oocytes (Miller and Beatty, 1969; Miller and Bakken, 1972; Hamkalo and Miller, 1973). Since its introduction, the technique has been modified repeatedly and it has been applied to the study of chromatin from a wide variety of chromosome types and cell types. Transcription complexes and chromatin conformations have been studied in embryonic tissues (Foe et al., 1976; Laird et al., 1976; McKnight and Miller, 1976, 1979; Busby and Bakken, 1979, 1980), in spermatocytes (Glatzer, 1975), in cultured cells (Miller and Bakken, 1972; Puvion-Dutilleul et al., 1977), in polytene chromosomes (Lamb and Daneholt, 1979), and in various kinds of DNA that have been injected into amphibian oocyte nuclei (Trendelenburg and Gurdon, 1978). Higher order structures of folded chromatin have been described in chick cells (Olins, 1977), in mouse cells (Rattner and Hamkalo, 1978a, b), and in amphibian oocyte nucleoli (Pruitt and Grainger, 1981). The technique has also been used in the study of translation units (polyribosomes) of bacteria (Miller et al., 1970), sea urchin embryos (Whiteley and Mizuno, 1981), insect salivary glands (Francke et al., 1982), and insect spermatocytes (A. H. Bakken and M. M. Lamb, unpublished).

The principle of the technique is the swelling and dispersal of nucleoplasmic chromosomal material in a low ionic environment followed by centrifugation of these three-dimensional structures on to a two-dimensional (flat) electron microscope grid surface. Such a preparation permits the investigator to trace long stretches of continuous chromatin fibres, identifying regions of actively transcribing 'genes' flanked by regions of silent or inactive chromatin. As one would expect, genes that produce a much used product such as ribosomal RNA are transcribed simultaneously by many RNA polymerase molecules (Figure 7.1(A)), whereas genes coding for structural proteins of the cell are usually transcribed at a lower rate and by fewer RNA polymerases (Figure 7.1(B) and (C)). Since initiation of transcription always begins at one end (5') of the gene and proceeds unidirectionally towards the 3' termination site, a transcription unit appears as a gradient of lateral, nascent (growing) ribonucleoprotein (RNP) fibrils, each of which is attached to the central DNP (DNA and protein) fibre by a putative RNA polymerase

---

(B) The beaded (nucleosomal) subunit structure of chromatin is easily visualized in this Miller spread from the nucleus of a 10 day fetal mouse red blood cell. The RNP fibrils are longer, less frequently initiated, and much more highly folded than in the ribosomal genes shown in (A). The arrow indicates the RNA polymerase molecule at the base of the RNP fibril. 8700×

(C) This TU was photographed in a Drosophila hydei primary spermatocyte. The RNP fibrils are thicker, shorter, and much less frequent than in (A) or (B). 8700×

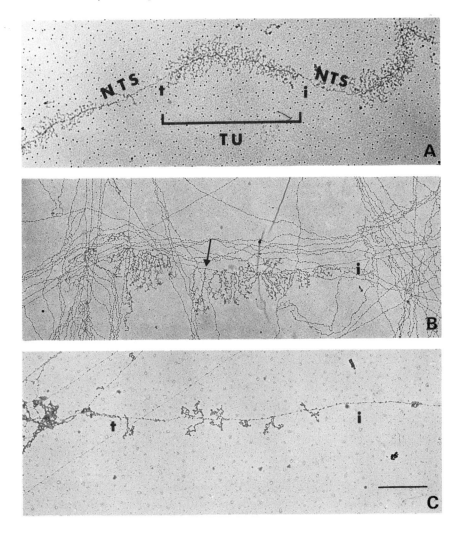

Figure 7.1 Representative transcription units (TUs) showing different RNA poly-merase densities. All preparations were made as described in Sections II.A, V.A, and V.B. They were stained with PTA, not shadowed, and photographed at 8700× with a Philips 300 EM (operated at 60 kV)

(A) Three tandemly repeated ribosomal RNA genes from a *Xenopus laevis* oocyte. The transcription unit (TU) encodes the 18 s, 5.8 s, and 28 s rRNAs which are transcribed together into a 40 s ribosomal precursor RNA molecule. Separating adjacent ribosomal TUs are non-transcribed spacer regions (NTS) of rDNA. The TU is composed of a gradient of rRNP fibrils of increasing length, from the 5′ initiation (i) end of the gene to the 3′ termination (t) end of the gene. 8700×

caption continued on page 162

complex (12.5–15 nm in diameter). The appearance of transcription units varies considerably in material from different cell types and even within the same preparation (Figure 7.2). Variations affect the overall length of transcription units, the length and conformation of the RNP fibrils, the spacing of the RNP fibrils, and the shape of the gradient of fibrils. Such variations result both from real differences between different genes (e.g. sequence-determined secondary structures in the RNA products and their packaging or folding with adherent proteins) and from variations in the preparation, i.e. the nature of the grid surface or the amount of residual salt or detergent or other compounds included in the dispersal medium. Such effects and how to handle them will be discussed in detail and examples provided subsequently. Suffice it to say here that, in order to visualize such transcription complexes, which in many somatic cell types are found on only 1–3% of the nuclear chromatin (the remaining 97–99% being transcriptionally inactive), one must unwind and/or disperse the chromatin fibres. Both low ionic strength solutions (essentially distilled water) and alkaline pH (8.5–9.0) are employed in this regard – and both are apt to remove some chromosomal proteins (including some of the histone H1), thus allowing the higher order chromatin fibres of the intact chromosome to unwind into the basic single-beaded strand (8–13 nm in diameter; Figure 7.1(B)). The tiny beads on this fibre have been called 'υ-bodies' (Olins and Olins, 1973, 1974; Woodcock, 1973), and are thought to represent nucleosomes or some modified form of nucleosomes. Chromatin and nucleosome structure has recently been reviewed by Worcel (1977) and by Georgiev *et al.* (1978). The effects of changes in ionic conditions on chromatin structure have been discussed by Thoma *et al.* (1979), Thoma and Koller (1981), Pruitt and Grainger (1981), and Labhart and Koller (1982).

Most 'Miller spreading' studies are based on the procedure of Miller and Bakken (1972), with modifications according to the type of material being examined. In what follows, methods are given for several systems, beginning with a detailed protocol for preparing and visualizing transcription units in

---

Figure 7.2 These two electron micrographs (A) and (B) were both taken from different preparations of 10 day fetal mouse red blood cells, prepared as described in Sections V.A and V.B, respectively. The grids were stained with PTA, but not shadowed, and photographs were taken at 3900×. For purposes of reproduction here, however, the magnification has had to be decreased to 2600×. Transcription units have variable numbers of RNA polymerases and variable lengths, and their RNP fibrils show quite different morphologies: a and aa are two different TUs with very similar moderate levels of activity and fairly long RNP fibrils; TU c shows a similar morphology in RNP fibrils, but they are more closely spaced on the chromatin fibre; the RNP fibrils in the TUs labelled b show a high degree of secondary structure and a much higher RNA : RNP packing ratio (more folding per unit length) than is seen in a or c. Some TUs are being read by only a single polymerase, s, or two polymerases, d. The most highly active transcription units, t, are ribosomal genes.

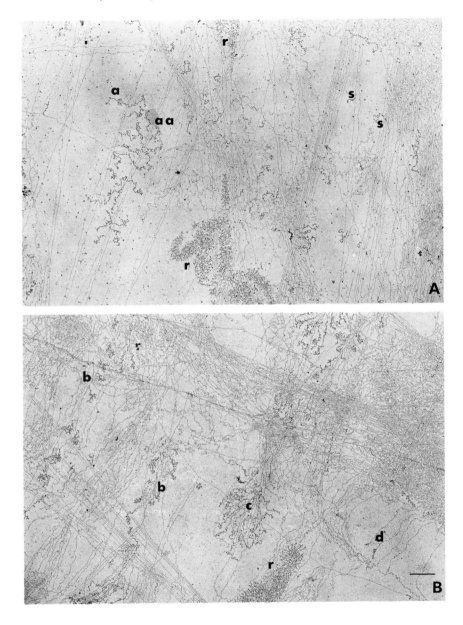

the nuclei of amphibian oocytes. The amphibian oocyte is an ideal model system. First, the oocyte and its nucleus (germinal vesicle, GV) arc sufficiently large to be handled with jeweller's forceps and the viewing aid of a dissecting microscope. Thus, one can prepare chromosomes and nucleoli using only mechanical disruption and dispersal to 'spread the genes'. A successful preparation of this sort, which contains thousands of active ribosomal genes and much lampbrush chromosome loop material, is aesthetically rewarding and also is a good standard test of the spreading conditions when one is adapting the method to a new system. A good oocyte preparation assures the investigator that the water supply is good, that the carbon-coated grid surface is sufficiently hydrophilic, and that there are no ribonuclease contaminants in any of the solutions or equipment. It also makes a good test system for trying out new ways of lysing cells and nuclei and seeing whether they modify or destroy transcription complexes.

I.   Visualizing transcription in oocytes

A.   Nucleolar (= ribosomal) DNA and lampbrush chromosomes

Before commencing a study of oocyte transcription by Miller technology, look first at Chapter 6 where information is given regarding choice of animals, methods, and equipment for looking at freshly isolated chromosomes and nucleoli with a phase contrast microscope, and methods for making permanent preparations for light microscopy.

(i)   Equipment needed

Details of items marked with an asterisk (*) are given in Section V.

   (a) Adapters for microcentrifugation chambers (see Figure 7.10)
   (b) Dissecting microscope (preferably of the stereozoom type) and black baseplate or a piece of black paper to cover baseplate
   (c) Centrifuge with a swing-out rotor, capable of up to 5000 rev/min
   (d) Coverglasses (circular, less than 20 mm in diameter)
   (e) Embryo cups or solid watchglasses with glass covers
   (f) Electron microscope grids, coated and glow discharged*
   (g) Filter papers (Whatman number 1, 9.0 cm in diameter)
   (h) Forceps, jeweller's number 5, straight and curved
   (i) Ice bucket with crushed ice
   (j) Lampbrush chromosome observation chambers (see Chapter 6)
   (k) Micropipettes (of the 'braking' type)*
   (l) Microcentrifugation chambers (MCC)* (see Figure 7.9)

(m) Petri dishes (new, plastic, 35 mm, 60 mm, or 9.5 cm for dispersal chamber)

(n) Very fine, sharp point, straight scissors

(o) Surgical needles (curved) number 23, and catgut suture (number 3/0), for use when it is required to obtain oocytes without killing the animal (see Chapter 6)

(p) Tungsten needles (see Chapter 6)

(q) Vaseline or soft paraffin grease

(r) Vellum or lint-free lens tissues

(ii)   Solutions needed

(a) Borate–NaOH buffer*

(b) 0.1% solution of MS222 (ethyl *m*-aminobenzoate methanesulphonic acid salt, or tricaine methane sulphonate; Sigma cat. no. E1626)

(c) 95% ethanol

(d) 4% formaldehyde–0.1 M sucrose solution*

(e) (For *Triturus*) A mixture of five parts of 0.1 M KCl and one part of 0.1 M NaCl, pH 7.0

(f) (For *Xenopus*) A mixture of five parts 0.07 M KCl and one part of 0.07 M NaCl, pH 7.6

(g) 4% phosphotungstic acid*

(h) pH 9 glass-distilled water*

(i) Glass-distilled water + 0.5% Photo-Flo 200 (Kodak), pH 9.0 with borate–NaOH buffer

(iii)   Procedure

The following protocol applies to material from *Xenopus* or *Triturus*. It is likely to be successful if applied to the oocytes of any amphibian. Two different salt concentrations have been given above for isolating *Xenopus* as opposed to *Triturus* germinal vesicles (GV). The nucleoplasm is much stiffer in *Xenopus* GVs, making it more difficult to disperse the lampbrush chromosomes and nucleoli. Whether or not this is a difficulty in the species of amphibian one chooses will become apparent in how easily the lampbrush chromosomes settle to the bottom of the observation chamber (Chapter 6). The rationale for various steps and hints for trouble-shooting are presented later.

(1) Anaesthetize the animal in MS222 (or in an ice bath) as directed in Chapter 6, Section III.A, or kill it if it is not required for subsequent experiments. The animal is anaesthetized when there is no longer a flexion response upon pinching the lower limb with forceps.

(2) With scissors and forceps, make a small incision (0.5–1.0 cm) in the skin, and then, similarly, in the abdominal wall away from the ventral mid-line. With one pair of forceps, hold the incision open, and with the other pair, pull out a piece of ovary. Cut it off with the scissors and place it in a dry, clean embryo cup. Seal a cover on to the embryo cup with vaseline and place it on ice. Stitch up the abdominal wall, then sew the skin closed (from two to six separate stitches). A Home Office Licence for experiments on live animals is required in the United Kingdom for operations of this kind.

(3) Remove a small bunch of oocytes from the sealed embryo cup and place them in 5 : 1 K/NaCl in another clean embryo cup.

(4) Place a drop of pH 9 glass-distilled water in a clean, plastic Petri dispersal dish. For *Xenopus* this drop should be 7–9 mm in diameter or about 400 μl. For *Triturus*, the drop need only be 5–7 mm in diameter or 160 μl volume.

(5) Using straight jeweller's forceps and a dissecting microscope (10–30×) with reflected light on a black surface, manually isolate two or three GVs in the manner described in Section III.B(iii) of Chapter 6, transferring each one to a second embryo cup with fresh K/NaCl. Use the straight micropipette for these transfers and always keep the GV in view through the dissecting microscope. Clean the GVs of any adherent yolk by repeatedly sucking each one in and out of the mouth-operated micropipette.* Rinse the pipette with distilled water to avoid yolk sticking to the inside.

(6) Using jeweller's forceps and/or tungsten needles, remove the nuclear envelopes quickly and carefully as described in Section III.B(iii) of Chapter 6. The nuclear contents of the GVs, which remain gelled and translucent for a short time, are then picked up in the tip of a smaller bore micropipette with a minimum of saline and transferred to the drop of pH 9 water. Let each GV drop out of the pipette into a different segment of the drop as you watch through the microscope.

(7) Cover the dish and allow the nuclear contents to disperse (*Xenopus* 15–20 min, *Triturus* 5–10 min).

(8) Meanwhile, prepare two wells in the microcentrifugation chamber(s) (see Figure 7.9). Fill each well with 0.1 M sucrose in 4% formaldehyde, pH 8.5.* Using a pair of forceps, take a carbon-coated, hydrophilic copper electron microscope grid*, rinse it with 95% ethanol, then sucrose–formaldehyde solution, then gently drop it into the well, allowing it to settle to the bottom with the carbon side uppermost. Remove all but one-quarter of the formaldehyde–sucrose solution.

(9) With a braking micropipette, slowly and gently pick up the drop of dispersed GVs with a circular motion around the drop to gently mix the solution as you fill the pipette. Now divide the drop between the two wells of the MCC(s), gently layering the nuclear contents on top of the sucrose–formaldehyde cushion.

(10) Place a round coverglass on top of the MCC. Blot away any excess fluid with filter paper (or bibulous paper). Put the chamber in a suitably adapted centrifuge tube or bucket (see Figure 7.10)*, and spin at 2500–3000*g* for 20 min.

(11) After centrifugation, remove the MCC from its holder with forceps or simply by inverting the bucket. With forceps, remove the coverglass and round up the meniscus with a drop of formaldehyde–sucrose solution. Invert the MCC. The grid will fall into the meniscus and can be removed easily with a pair of curved jeweller's forceps. If a grid sticks to the bottom of the MCC, it can be dislodged with a very fine tungsten needle.

(12) Holding the grid with the curved forceps, rinse it immediately in pH 9 water and then in pH 9 water + 0.5% Photo-Flo 200 solution, for 7 s each.

(13) Draw off the excess fluid from between the forceps and from the grid by pulling its edge across a piece of vellum tissue. Air dry.

(14) To stain the grids, either float them face down on a drop of 1% phosphotungstic acid* on a clean piece of parafilm (or a clean Petri dish) or, holding each grid in forceps, immerse it in the stain in the well of a microtitre plate. Stain for 30 s. Then rinse twice in 95% ethanol and once in pH 9 water + Photo-Flo 200. Again dry the grid by drawing off the fluid with vellum tissue. To enhance contrast of the DNP and RNP, the grids can be stained also with 1% uranyl acetate and/or can be rotary shadowed with platinum/palladium in a vacuum coating unit.

(15) The preparations are now ready for viewing in an electron microscope.

B. Alternative methods for lampbrush chromosomes

Due to the large size and extended conformation of amphibian lampbrush chromosomes, they are easily broken during the procedures described above. Thus, although one will see long pieces of DNP with lots of RNA polymerases and long, lampbrush-type RNP fibrils in such preparations, it is exceedingly rare to see a whole, intact lampbrush loop. A simple modification of the general procedure reduces the handling of the GVs and their lampbrush chromosomes, thereby minimizing breakage of chromosomes and lampbrush loops.

(i) Additional solutions and equipment

(a) Glass-distilled water + 0.025% 'Joy' (Joy lemon-fresh washing up liquid, Proctor and Gamble Co., Cincinnati, Ohio, USA), pH 9 with borate–NaOH buffer

b) Moisture chamber

(ii)   Procedure

(1) Prepare an MCC. Remove half the formaldehyde–sucrose solution and add the 'Joy' solution to make two distinct layers.

(2) Isolate two GVs; wash off the adhering yolk.

(3) Using the braking micropipette, transfer each GV through several rinses of pH 9 water in embryo cups (keeping the GV in view through the microscope at all times), and then place it in the upper layer of one of the wells of the MCC. The nucleus will rupture and the nuclear contents will disappear from view as soon as they come in contact with the 0.025% 'Joy' in the chamber.

(4) Gently place the MCC in a moisture chamber and leave it undisturbed for 15 min to allow the nuclear contents to disperse.

(5) Add a coverglass and centrifuge at 2500–3000g for 20 min.

(6) Remove the grid from the well, rinse, dry, and stain.

This method may seem simpler, due to the reduced handling of the nuclei. However, the use of detergents such as 'Joy' may reduce the amount of material which 'sticks' to the grid surface. Also, dispersal is not always as complete as in the 'drop', which has a wider diameter than the MCC well. Thirdly, the detergent tends to remove chromosomal proteins, thereby reducing contrast of the DNP and the RNP against the carbon support film. Rotary shadowing* of the preparation with 80% platinum/20% palladium metals is strongly recommended.

II.   Electron microscope images of transcription in oocytes

Viewing 'Miller spreads' in the electron microscope (EM) is quite different from viewing sectioned material. In this case one is looking for fine strands of macromolecules against a somewhat grainy background of carbon support film. However, once a cluster of active ribosomal RNA genes have been observed with their characteristic feather-like or 'Christmas-tree' shape (due to the short to long gradient of RNP fibrils), thereafter they will always be immediately recognizable, even in a poorly spread mass of inactive chromatin.

Begin scanning the grid, using the binoculars, at a magnification of 8000–10 000×. Once a nucleolus (which may occupy from a quarter to a half of a 400 mesh grid square) has been found, look at the image just through the viewing window (see Figure 7.3 for a low magnification view of a nucleolus). Note the characteristic shadow on the grid surface and the dots and dashes of alternating transcription units as compared with non-transcribed spacer regions between them.

Figures 7.4–7.6 show examples of the kinds of transcriptional activity likely to be encountered in Miller spreads from amphibian oocytes.

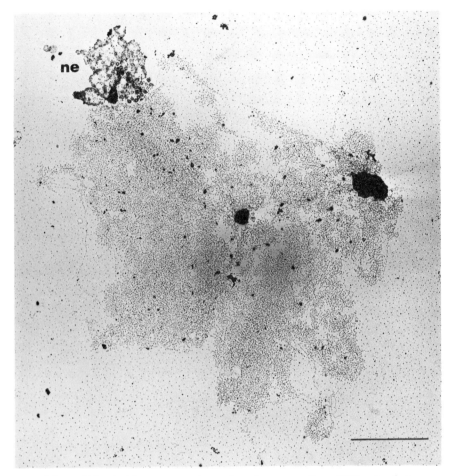

Figure 7.3 A *Xenopus* oocyte nucleus has more than 1000 extrachromosomal nucleoli due to amplification of the rDNA during pachytene. One such nucleolus, prepared as described in Section II.A, is shown here. It was stained with PTA, but not shadowed, and photographed at 3900×. Several individual ribosomal RNA TUs can be distinguished at the periphery. A piece of nuclear envelope (NE) is seen in the upper left corner, with a number of nuclear pores. The scale bar represents 2.5 μm

Figure 7.4 shows a circle of tandemly repeated 28 s + 5.8 s + 18 s ribosomal DNA transcription units that form the basis of the large number of extrachromosomal nucleoli produced by gene amplification in amphibian oocytes (see review by Macgregor, 1972). The fine scale morphology of transcription units of this kind is typical of active ribosomal DNA from a wide variety of eukaryotes. In Miller spreads, active ribosomal DNA has three strongly distinguishing features. First, the transcription units are always

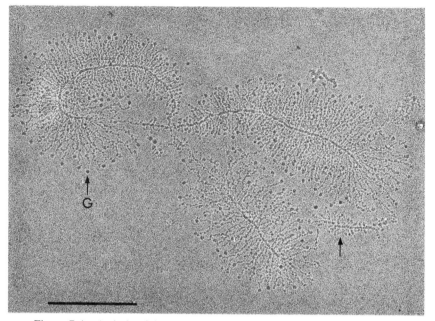

Figure 7.4   A circle of three extrachromosomal transcription units coding for the 18 s+5.8 s+28 s ribosomal RNA of *Xenopus laevis*. Each tandemly repeated unit is separated from its neighbours by a variable length of transcriptionally inactive chromatin representing the non-transcribed spacer segments of the ribosomal gene complex. Occasionally these spacer regions show small transcription units ('prelude complexes'), one of which is indicated by the arrow in the picture. A conspicuous granule (G) is always present at the distal end of ribosomal RNP transcripts. The scale bar represents 1 μm

tandemly repeated, although in some protozoa there may be as few as two units in any one tandem array (Gall *et al.*, 1977). Secondly, the nascent transcripts (RNP fibrils) always have a terminal bead or granule at their outer ends (labelled G in Figure 7.4), which first appears one-quarter to one-third of the way down the gradient. Third, ribosomal RNA genes are nearly always maximally packed with RNA polymerases. Most transcription units, particularly in somatic cells, have many fewer polymerases on them.

Lampbrush chromosomes, like ribosomal RNA genes, often show a very high rate of RNA synthesis. Figures 7.5 and 7.6 show two typical transcription units from lampbrush chromosomes of *Xenopus laevis*. In Figure 7.5 the transcripts are closely spaced and form a clearly defined gradient from left to right along the length of the transcription unit. In figure 7.6 the RNP fibrils are more widely and irregularly spaced and the chromatin axis has the beaded appearance typical of that found in transcriptionally inactive regions of chromatin. Also it is evident in Figure 7.6 that the RNP fibrils are complex

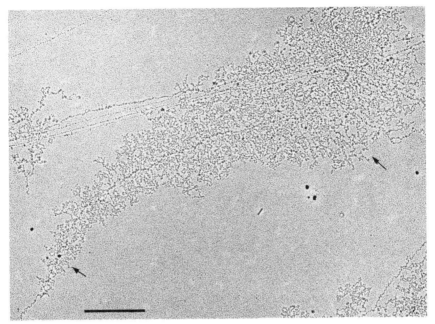

Figure 7.5   A typical transcription unit on a lampbrush chromosome of *Xenopus laevis* spread by the method described in Section I.B. The RNP fibrils (arrows) are closely packed and form a gradient of increasing length from start to finish of the transcription unit. Scale bar represents 1 μm

structures showing a high degree of packaging and folding. As will be noted later in this chapter, transcription units can present a great variety in their length, the frequency of polymerases in transit, the degree and specificity of folding patterns in the nascent transcripts and the frequency with which they are found in a particular cell type. Several laboratories have used these parameters to characterize changes in transcription patterns during development for this method permits the investigator to quantify the number of polymerases per gene at a particular moment in time. In contrast, biochemical methods for quantifying new RNA synthesis can only measure average rates of synthesis per gene, whether for the total number of genes active at that time or on the assumption that all copies of a specific repeated gene are active at that time.

III.   Rationale and comments on methods for germinal vesicles

Triturus oocytes keep well in a dry, sealed dish at 4–5 °C for 4 days. Thus for classes, or repeated experimentation, the animal can be killed, both ovaries removed, and small pieces cut off by each student as required.

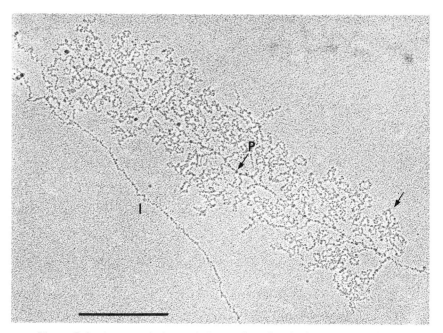

Figure 7.6   A transcription unit from a lampbrush chromosome of *Xenopus laevis*. Here the RNP transcripts are widely and irregularly spaced and consist of beaded fibres that are folded to form bush-like structures (arrow). Between transcripts the chromatin axis has a distinct nucleosomal appearance like that of adjacent transcriptionally inactive chromatin (I). The RNA polymerase complex (P) at the base of each RNP transcript is larger than the nucleosomes on the chromatin axis. Scale bar represents
1 μm

Xenopus oocytes may be kept in modified Barth's solution* at 4–18 °C for 1–2 days, but are best when used the same day they are taken from the animal.

When working with single nuclei, even large ones like amphibian oocyte GVs, many beginners have trouble with 'losing the nucleus' somewhere during the procedure. This can easily be avoided. Always keep the nucleus in view through the dissecting microscope. When cleaning the nucleus or transferring it, keep one hand on the micropipette and the other on the embryo cup. The GV is quite sturdy. When cleaning off the GV, by pulling it in and out of the micropipette, one can bounce it off the bottom of the dish without injury. Do not, however, allow the GV to touch an air/water interface, either at the surface of the embryo cup or in the pipette, for it will rupture immediately. Ensure that the micropipette draws slowly enough that full control of the fluid and the GV is maintained. Do not allow the fluid to rise high enough to enter the constriction in the pipette. If it does, rinse the pipette with distilled water and dry it carefully in a bunsen.

Always keep a beaker of distilled water next to the microscope and rinse the micropipette frequently. Yolk granules, etc., appear small under the dissecting microscope, but they are huge and will obscure the chromatin and transcription units when they are viewed in the preparation by EM.

Gather all equipment together before beginning these preparations. The EM grids can be coated with support film 1–3 days ahead of time and may keep 1–2 weeks.

Glow discharge* just the number of grids required at the beginning of each day (or make them hydrophilic with ethanol as the preparation is being made).

It is important actually to remove the nuclear envelope, not just prick it or tear it open, for once the lampbrush chromosomes start oozing out of a hole they become ferociously sticky, attaching to the forceps, the embryo cup, the micropipette, etc., and subsequent dispersal is poor.

The modified procedure in which the intact GV is placed in a 'Joy' detergent solution in the MCC for lysis and dispersal is a superb method for preserving lampbrush chromosomes intact. It has been used also by the authors when handling only a few cells of a given type, e.g. mouse oocytes (Bakken, 1976), where it is critical not to lose any of the material. There is a disadvantage, however, in that transcriptionally active chromatin from whole nuclei sometimes lands on the surface of the grid as large aggregates of undispersed material.

When transferring whole nuclei to the MCC, as opposed to disrupted ones, the use of 'Joy' detergent is essential for destruction of the nuclear membrane and to promote dispersal and spreading of the chromosomal and nucleolar material. However, 'Joy' removes chromosomal proteins and so its concentration should be kept to the minimum required for removal of the nuclear membrane.

With material from some species, and particularly with oocyte nuclei from crested newts, 'Joy' need not be used. The oocyte nuclei swell rapidly in the pH 9 water and can easily and gently be mechanically ruptured by letting them swell until they become a little too large for the micropipette. The swollen nucleus will then have its membrane stripped away when it is sucked up into the pipette and the contents can be carefully emptied into the MCC without serious disruption and allowed to disperse.

Staining with PTA rarely imparts sufficient contrast to detergent-treated chromatin or transcription units and rotary shadowing in addition to staining is recommended. Even so, the image contrast after shadowing is not high when seen on the screen of the electron microscope, although the thinner the support film, the better the contrast. With good film and some practice, one can scan amphibian oocyte grids without binoculars at 3000–4000× for the nucleoli are quite distinctive.

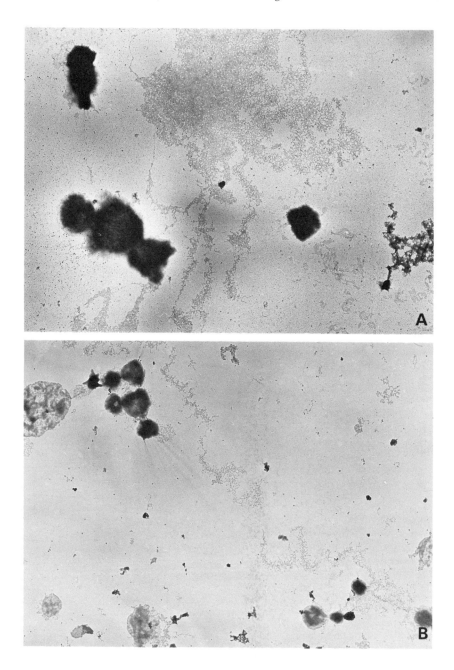

Several parameters can affect the spreading of oocyte transcription units. Too much salt transferred with the nuclei to the dispersal drop will prevent the nucleoli from unwinding and may cause the RNP fibrils to remain 'matted' together even at the edges of the nucleoli (Figure 7.7(A)). Even if the dispersal has been sufficient, the genes can be spread but still appear matted or precipitated (Figure 7.7(B)). This is usually the result of a hydrophobic carbon surface. Remember to test the grids for their hydrophobic character (see Section V) and perhaps glow discharge longer. The pH of the solutions also has a significant effect on spreading. If either the pH 9 water or the formaldehyde–sucrose solution have dropped in pH to around 7, the chromatin will not disperse well.

IV. Modification of the technique for other cell types

The procedure for dealing with embryonic and somatic cells is basically the same as has been given in previous sections. The main difference is that the cell nuclei are too small to isolate manually and so the cells *and* nuclei need to be lysed with detergent. Different cell types require slightly different treatments with detergent (time and concentration). Two manipulations will help the investigator in controlling these parameters. First, it is helpful to do the lysis and dispersal in a small volume vessel, since the detergent decreases surface tension and prevents the dispersal medium from standing up in a drop. The wells of a microtitre plate or small glass test tubes make excellent dispersal chambers. Secondly, the investigator should monitor the progress of the lysis and dispersal by taking small aliquots (at frequent intervals) and checking the state of the nuclear rupture in a lampbrush chromosome observation chamber under the phase contrast microscope. At first one will see the nuclear (and cytoplasmic) contents 'dissolve' and become a uniform (smooth) grey. Then a hole appears in the nuclear envelope and a small cloud of nucleoplasm extrudes from it. Finally, one will see only small remnants of nuclear envelope and a large diffuse cloud of nuclear material. This material is ready for centrifugation on to an EM grid. Because detergents remove proteins from the chromatin and the transcription complexes, it is important to use the smallest amount required and for the least amount of time. However, good dispersal may be achieved with longer dispersal

Figure 7.7 Both photographs were taken at 5600× after PTA staining, but no shadowing. Magnification has had to be reduced to 4200× for purposes of reproduction. *Xenopus* oocyte nuclei were prepared as in Section IIA. However, in (A), too much salt was transferred with the nuclei to the dispersal drop, so the nucleolus shown is not very well unwound and the RNP fibrils are 'matted' or 'precipitated' on each gene. (B) A nucleolus that is well dispersed, but the grid surface was still too hydrophobic, giving the rRNA genes a 'matted' or 'precipitated' appearance. Both photographs contain darkly staining membranous and non-chromatin material

times. Thus, it is sometimes useful to treat the cells and nuclei with a small volume of detergent solution, monitor the nuclear rupture and then add pH 9 water to dilute the detergent concentration tenfold or more during the dispersal time. The dilution of the detergent will also increase by up to tenfold the amount of material that sticks to the grid.

A. Embryonic cells

This method was first applied to cells from *Drosophila melanogaster* (McKnight and Miller, 1976). It has since been successfully applied to material from several other species. The protocol given here is designed to be generally applicable. Naturally, each individual investigator may need to introduce small modifications according to the system in use and the questions that are being asked of it.

(1) Prepare an MCC in the usual way but load it only with the formaldehyde–sucrose solution and a coated grid. Do not put pH 9 water on top of the formaldehyde–sucrose.

(2) Isolate, stage, and rinse the embryos in an appropriate saline for the species in question. A modified Locke's insect Ringer has been recommended for *Drosophila**.

(3) Mince a whole embryo or a part of one as required with two pairs of watchmaker's forceps in a 100 µl droplet of 0.05% 'Joy' adjusted to pH 9 with NaOH–borate buffer* to form as near a single cell suspension as possible. This operation is best carried out with the droplet on the surface of a siliconized coverglass.

(4) Stir the droplet gently for 1 min with the point of a tungsten needle.

(5) Add to the droplet 100 µl of a solution containing formaldehyde–sucrose solution, pH 8.5*.

(6) Leave to stand for 5 min and then very carefully layer the whole of the 200 µl droplet on top of the formaldehyde–sucrose layer in the MCC.

(7) Centrifuge and process exactly as described in previous sections.

If no transcription units can be found in the EM preparations, there could be a high level of endogenous ribonucleases at work. Adding yeast tRNA (Sigma) at 100 µg/ml to the pH 9 water dispersal medium may be effective in competing out the RNase activity.

B. Tissue culture cells

The method described in this section was introduced by Miller and Bakken (1972) for the study of chromatin and transcription units from nuclei of HeLa cells. It has been modified by several workers for application to other kinds of cells in tissue culture (Puvion-Dutilleul *et al.*, 1977; Harper and Puvion-Dutilleul, 1979).

(1) Culture the cells until they are in the log phase of growth (10^6–5×10^6 cells/ml).

(2) For cells in spinner cultures, gently pellet the cells (2000 rev/min, 3 min), in a conical centrifuge tube. (For cells grown on plastic surfaces, trypsinize them off and wash once with media and twice with a balanced salt solution.)

3) Add an equal volume of 0.1–0.5% 'Joy', pH 9*, to lyse the cells and nuclei. Monitor the lysis in aliquots with the phase contrast microscope. When most nuclei are ruptured, add 10 volumes of pH 9 water and allow to disperse. Check dispersal with a phase contrast microscope.

(4) When nuclei are dispersed, place the dispersal vessel on ice and add 0.1 volume of formaldehyde–sucrose solution to stabilize the chromatin while several EM grids are prepared.

(5) Layer the mixture into the well of an MCC that is already one-quarter filled with formaldehyde–0.1 M sucrose solution.

(6) Centrifuge and process for electron microscopy as described previously.

(7) Immediately, check the preparation in the EM for concentration of material. (Sometimes this can be accomplished by viewing the EM grid in a dry lampbrush chromosome observation chamber under the phase contrast microscope with a 40× objective magnification.)

(8) Make more EM grids, adjusting the concentration as necessary.

(9) Stain with PTA (and uranyl acetate if desirable), check the grids in the EM, and rotary shadow if necessary.

It is likely that spreading techniques for tissue culture cells will need to be varied according to the kinds of cells being used. The most likely variables are the concentration of detergent needed for cell lysis, the time needed for dispersal of the chromatin, and the number of cells that must be used to make useful preparations. Monitoring the nuclear lysate with the phase contrast microscope will greatly aid one in defining the best conditions for a particular cell type.

C. Transcription of DNA injected into oocyte nuclei

Transcriptionally active cloned ribosomal genes that had been injected into *Xenopus* oocytes were first seen in Miller spreads by Trendelenburg and Gurdon (1978). The approach is of special value since it permits analysis of the size and transcriptive activity of specific genes on an individual basis.

Determining polymerase density on each copy of the gene is quite important for defining the effects of different promoter loci.

Either of the methods outlined in Sections I.A and I.B can be used to visualize injected material, the only recommended modification being that preparations should be centrifuged at 10–15 °C between 10 000g and 20 000g for half an hour so as to increase the number of injected molecules that reach the surface of the grid. Some investigators, notably Trendelenburg *et al.* (1980), recommend the addition of 0.02% Sarkosyl NL30 (Ciba-Geigy) to the pH 9 water to break up aggregates of plasmids or other small injected molecules. However, if the whole oocyte nucleus technique described in Section I.B is used, then the 'Joy' present in the MCC has the same effect as Sarkosyl. Bakken *et al.* (1982a, b) did not use detergent and found some interesting chromatin conformations in the non-transcribed spacer DNA.

At the time of writing this chapter, the usefulness of the injection–Miller spreading technique is limited by three factors. First, of the vast number of genes that are normally injected into an oocyte nucleus in an 'average' experiment, perhaps 10^9 copies, it is rare to find more than about one hundred active ones on a single grid. Screening grids for good material can therefore be very time consuming. The second problem concerns the choice of material to be injected and spread. For meaningful and accurate analysis relatively large genes of perhaps 3000–4000 base pairs are needed. While there are many cloned gene sequences of that size now available, not many of them are accurately transcribed in the environment of the oocyte nucleus. This particularly applies to genes transcribed by RNA polymerase II (Wickens *et al.*, 1980). The third consideration is that data on the activity of injected material can very often be obtained much more accurately and quickly by biochemical methods. Nevertheless, Miller spreading in conjunction with micro-injection does have substantial possibilities, as has recently been well shown by Bakken *et al.* (1982a, b), who have successfully spread cloned ribosomal genes and restriction fragments of them, combining most effectively electron microscopy and biochemistry to map initiation and termination signals for transcription in these cloned DNAs.

The transcription of recombinant DNA after injection into oocytes has recently been reviewed by Wickens and Laskey (1981) and by Kressman and Birnstiel (1980).

V. Solutions and equipment for Miller spreading

(a) *Sodium hydroxide–borate buffer*. Dissolve 3.1 g of boric acid in 250 ml of glass-distilled water to form an approximately 0.2 M solution. Adjust the pH of the solution to around 9.5 by adding about 30 ml of 1 M NaOH. Keep the buffer in the refrigerator but not for more than 1 month. This solution is used for adjusting the pH of most other solutions. One can substitute

Mallinckrodt pH 10 'buffAR'. But most other buffer standards contain germicidal or fungicidal agents that are detrimental to 'good spreading'.

(b) *Formaldehyde sucrose.* This solution forms the lower layer in the MCC, the formaldehyde acting as a fixative in the histological sense and the sucrose as a density cushion to trap small nucleoplasmic (and in some preparations cytoplasmic) particles that clutter the grid surface. When nuclear lysis and dispersal are performed in the MCC, the cushion regulates the treatment of the chromatin after it has been added to pH 9 water forming the upper layer in the MCC. Essentially, the chromatin is retained in the pH 9 water and therefore fully exposed to its dispersive effect until such time as the MCC is centrifuged. Only then does the chromatin pass into the sucrose-formaldehyde and down onto the grid.

Bring a little more than 100 ml of glass-distilled water to a pH of between 8.5 and 9 by adding a little borate–NaOH buffer. Weigh out 4 g of paraformaldehyde powder and add pH 8.5–9 water to a final volume of 100 ml. Stir and warm until the paraformaldehyde is dissolved. Do not boil the solution and do the whole operation in a fume cupboard. When the solution is clear, allow it to cool and then add 3.4 g of sucrose (Grade 1, Sigma cat. no. S9378) to make an approximately 0.1 M solution. Filter the solution through a nitrocellulose bacterial filter, pore size 0.45 μm, and store it in a refrigerator. The solution can be kept for up to 1 week, but its pH will fall. Small aliquots should therefore be removed as needed and readjusted to pH 8.5 immediately before use.

(c) *pH 9 water.* Bring glass-distilled water to pH 9 by the dropwise addition of NaOH–borate buffer. Check and adjust the pH of the water every 2–3 h. Keep in a covered beaker filled to the top. Make up fresh pH 9 water before every session of Miller spreading. Alternatives to NaOH–borate for pH adjustment are 10 mM $NaHCO_3$ or Mallinckrodt pH 10 'buffAR'. In the latter case about 70 μl of 'buffAR' are needed for 50 ml of water. For extra long dispersal times (30–40 min for material that is highly compacted or hard to spread) one can use pH 9 water that is supplemented with 5 mM sodium borate. This will maintain the pH of the dispersal drop.

Glass-distilled water that is stored in some kinds of plastic bottles may tend to accumulate noxious compounds from the plastic that prevent good spreading. This will be quite apparent as when a GV is placed in the pH 9 water and observed through the dissecting microscope the nucleus appears white due to the denaturation of the proteins. Thus, store the distilled water and other solutions in glass bottles.

(d) *Detergents.* Add 'Joy' or Photo-Flo to pH 9 water, if appropriate, immediately before use. Then readjust the pH if necessary with NaOH–borate. Keep special stock bottles of these detergents for Miller spreading only, and dip into them with very clean pipettes only. Detergents can become contaminated with microorganisms and fungi. It may be wise to filter (0.45

µm pore size nitrocellulose) the working solution if there are problems with bacteria on the grids.

(e) *Modified Locke's insect Ringer*. According to McKnight and Miller (1976) this medium is composed as follows: NaCl, 9 g; KCl, 0.42 g; $CaCl_2$, 0.25 g; $NaHCO_3$, 0.20 g; make up to 1 l with distilled water.

(f) *Modified Barth's medium*. Gurdon (1974) describes the composition of this medium as follows: NaCl, 5.13 g; KCl, 0.075 g; $NaHCO_3$, 0.20 g; $MgSO_4.7H_2O$, 0.20 g; $Ca(NO_3)_2.4H_2O$, 0.08 g; $CaCl_2.6H_2O$, 0.09 g; benzyl penicillin, 0.01 g; streptomycin sulphate, 0.91 g; Tris-HCl,pH 7.6, 0.91 g; make up to 1 l with distilled water; adjust to pH 7.6 if necessary with HCl or NaOH.

(g) *Phosphotungstic acid stain (PTA)*. Add 0.2 g phosphotungstic acid (Polaron Equipment Ltd cat. no. NC 3009) to 5 ml of glass-distilled water to give a 4% stock solution. Filter through a nitrocellulose filter, pore size 0.45 µm, and store in the refrigerator. Do not adjust the pH. Dilute one part of the stock with three parts of 95% ethanol immediately before use.

PTA stains proteins and is adequate staining for most non-detergent-treated chromatin. Staining longer than 30 s appears to decrease the contrast of the chromatin in the EM and probably removes proteins. The grids may be counterstained with uranyl acetate (prepared in the same way, i.e. 4% aqueous stain diluted 1 : 3 with 95% ethanol), which stains the nucleic acid. In addition, the preparations may be rotary shadowed with 80% platinum/20% palladium metals at a 7–8° angle if desired.

(h) *Braking micropipettes*. The design of braking micropipettes used in handling germinal vesicles and chromatin is an important part of the technique.

One effective design is illustrated in Figure 7.8(A). The pipettes should be made of soda glass capillary tubing with an internal diameter of about 1.3 mm and a wall thickness of about 0.3 mm. The internal diameter of the finished drawn tip must be varied according to the size of the nuclei being handled and the method of Miller spreading being used. In this design the braking effect is obtained by the constriction produced when the pipette is bent at right angles. The pipette is operated with a piece of rubber tubing and a plastic mouthpiece.

An alternative method of making braking pipettes is described in Stone and Cameron (1964). A piece of glass tubing, 15–20cm long, 3–4 mm outside diameter, 2 mm wall diameter, is heated to produce a braking constriction near one end, and the tip of the tube is bent and drawn out in a separate operation (Figure 7.8(B)).

The surfaces of the micropipettes may be siliconized by dipping in 'Repelcote' (Hopkin and Williams cat. no. 996270) or an equivalent silicone solution. Coated glass should be baked in a hot oven before use to ensure that the silicone coating is firmly attached to the glass surface.

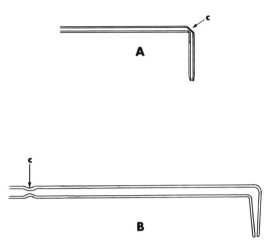

Figure 7.8   Two designs for braking pipettes suitable for picking up and transferring cell nuclei or nuclear contents. (A) A pipette made from capillary tubing. The width of the mouth of the pipette should be made to suit the kinds and sizes of nuclei that are to be handled. The braking effect in this design is produced by a constriction (C) that forms at the right-angle bend. (B) A pipette constructed from 3–4 mm glass tubing. Two steps are involved in making this kind of pipette. First a braking constriction (C) is produced, and then the tip of the pipette is drawn out to the appropriate width. In this diagram the pipette is shown with a bent tip. It will function equally well if left straight

Pipettes should be washed between each transfer of nuclei in a beaker of distilled water kept by the microscope. Micropipettes are reused for months or years in the laboratory. Problems with ribonuclease contamination of micropipettes or MCCs, etc., are rare. The most frequent source of ribonuclease contamination is from wiping the tips of the jeweller's forceps with bare fingers. However, to be safe some workers recommend that all glassware used in Miller spreading should be cleaned in chromic acid and baked at 180 °C before use.

(i) *The MCC.* The chamber is constructed from Perspex or 'Plexiglass'. A recommended design and dimensions are shown in Figure 7.9. The wells must be milled so that their sides are smooth and the bottom flat so as not to puncture the carbon support film on the EM grid. The MCC can be constructed by milling to within about 1 mm of the bottom of the plastic disk, so leaving a solid plastic base, or drilling right through the disk and sealing a thick coverglass over the bottom with a strong resin adhesive.

For high speed centrifugations, as when one is studying the transcription of recombinant plasmids or other DNAs micro-injected into oocyte GVs, the

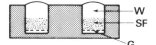

Figure 7.9   Scale diagram of a microcentrifugation chamber (MCC) as seen in side section. The MCC has two wells. Each is shown filled with pH 9 water (W) and sucrose–formalin (SF), and there is an electron microscope grid (G) at the bottom of the well. The shaded material represents the Perspex (Plexiglass) of the MCC

application of two or three coverglasses interspersed with coatings of epoxy glue is recommended.

One can use MCCs of different heights which thus hold different volumes. Originally, Miller used MCCs with a well of 6 mm in depth and 4.5 mm in diameter for viewing *Triturus* GV transcription. Bakken uses MCCs with a depth of 12 mm for *Xenopus* GVs because these nuclei disperse better in larger volumes of pH 9 water. Hill also uses tall chambers to allow for good dispersal of the lampbrush chromosomes in 'Joy' solution within the MCC. In addition, taller chambers permit one to vary the thickness of the sucrose cushion. Thus if one is studying a yolky fish oocyte or insect embryo, the sucrose layer can be increased both in molarity to 0.3 M, 0.5 M, or more and in height within the MCC.

The outer diameter of the MCC can be modified to fit any size bucket for the swing-out rotor available to the investigator. It is important to support the MCCs on a flat surface at the bottom of the bucket. A simple way of doing this is to place a slightly tapered rubber bung (stopper) in the bottom of the bucket and the MCC on the top of that. The MCCs slide down onto the bung with ease if they just fit the bucket. To remove the MCCs after centrifugation, simply invert the bucket. The chromatin has already stuck to the grid surface and will not be disturbed by this procedure. A more complex system, but one that is generally more satisfactory in the long term, is shown in Figure 7.10. An Araldite plug is cast in a plastic centrifuge tube and the top surface of the Araldite is then flattened and shaped with a lathe to accommodate an MCC. A hole is bored in the bottom of the centrifuge tube such that the Araldite plug can be pushed to the top of the tube for loading and unloading of the MCC.

MCCs should be cleaned in warm soapy water, using a pipe cleaner to reach the inside of the wells. They should be rinsed thoroughly in distilled water and air dried. Do not clean them in ethanol and do not use very hot water; both are liable to damage or distort the MCC.

(j) *Coated grids for Miller spreads.* The quality of the support film coating the EM grids is very important for Miller spreads. Since one is viewing individual strands of DNP and RNP, the differences in contrast between the chromatin and RNP fibrils and in contrast between the chromatin and RNP

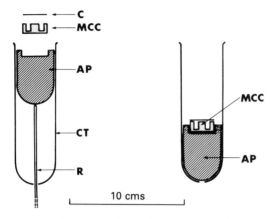

Figure 7.10 Scale diagrams of an adapter system for accommodating MCCs in a 30 ml plastic centrifuge tube. The right-hand diagram shows the Araldite plug (AP) and MCC in position in the tube and ready to centrifuge. The left-hand diagram shows the Araldite plug pushed to the top of the tube for loading or removal of the MCC. C = coverglass, CT = centrifuge tube, R = rod or seeker used for pushing the araldite plug to the top of the tube to facilitate loading or removal of the MCC

fibrils and the 'background' of the support film must be optimized. This means, the thinner the support film the better.

Grids of 300 and 400 mesh are recommended. They can be purchased with one shiny side and one dull side. Most laboratories always put the carbon film on the shiny side so that everyone knows which side is 'up'. The grids may be coated with carbon alone, carbon and Parlodion (nitrocellulose), or carbon and Formvar. Carbon alone (as evaporated on to mica, floated on to water, and then lowered on to copper grids), will give the strongest, thinnest film. Plastic films of Parlodion or Formvar (made with 0.1–1.0% plastic in solvent), must be stabilized with a thin layer of carbon and are thus thicker. Parlodion appears to promote better spreading of the RNP fibrils than Formvar. The reason for this difference is unknown. Methods for coating grids are given in most textbooks on electron microscope technique and will not be covered here (or see Bakken and Hamkalo, 1978).

Coated grids must be made hydrophilic before they are used for Miller spreading. This is best done immediately before use by glow discharging them in a vacuum coating unit (vacuum evaporator) at 0.1 torr for 1–2 min with a current of 10–15 A. Alternatively, carbon-coated or carbon–Parlodion-coated – but *not* Formvar–carbon-coated – grids can be made hydrophilic by agitating them in 95% ethanol for 2–3 min and then rinsing thoroughly in pH 9 water immediately before placing them in the MCC. To test whether a grid is fully hydrophilic, rinse the grid with pH 9 water. If the carbon side of the grid remains covered to the edges with water, it is hydrophilic. If the water 'beads up' towards the centre, the grid surface is still hydrophobic and should be rinsed longer in ethanol. Different batches of carbon film will be more or less hydrophobic, so this test will allow one to adjust the ethanol washing time or glow discharging time as necessary.

(k) *Rotary shadowing of grids*. Rotary shadowing of the grids is useful in several circumstances. It will enhance contrast of Miller spreads when (1) the chromatin is very thin due to removal of proteins during cell and nuclear lysis with detergents in the dispersal medium or due to over-staining, and (2) when the support film is thicker than preferable. Shadowing does make scanning of the grids a little easier on the eye. But one should note that the dimensions of structures in the preparation will be increased by the shadowing and thus accurate measurements may not be obtained.

The procedure for rotary shadowing will naturally vary according to the kind of vacuum coating equipment being used. However, the following guidelines apply to most systems and will form the basis for a method that should prove to be effective.

Use a V-shaped filament of 1 mm thick tungsten wire. Wind around the V of the filament 2 cms of 0.2 mm diameter 80% platinum/20% palladium wire. Position the filament 12 cm from the surface of the rotating table such that a line joining the tip of the V-filament and the centre of the table is at $7°$ to the horizontal. Attach the grids to a piece of microscope slide with double-sided adhesive tape and place the slide on the rotating table such that the grids lie quite flat with the preparation side uppermost. Arrange for the table to rotate at 20–30 rev/min. Evaporate the platinum/palladium at 10^{-4}–10^{-5} torr over a period of 30–60 s. Allow the filament to cool for 5 min before admitting air to the chamber.

VI. Analysing chromatin spreads and transcription units

It is best to take low power electron micrographs of stretches of chromatin which can then be enlarged several-fold when printing. Taking photographs at high magnifications tends to amplify the 'graininess' of the carbon support film and blur the edges of the DNP and RNP. Detailed morphology of chromatin and transcription units is best appreciated by serially photographing long stretches of chromatin and then constructing montage prints. In constructing montages involving many prints the use of resin-coated photographic paper is recommended since it can be processed rapidly and does not require glazing.

Electron micrographs of Miller spreads are ideal for quantitative analysis. Discussions on methods of quantitation are given by Busby and Bakken (1979), Chooi (1979), Foe *et al.* (1976), Glatzer (1979, 1980), Hill (1979), Hill and Macgregor (1980), Laird *et al.* (1976), Laird and Chooi (1976), McKnight and Miller (1979), and Puvion-Dutilleul *et al.* (1978).

Inactive chromatin in Miller spreads normally appears as a beaded fibre 8–13 nm thick (Figure 7.6). The beads are considered to correspond to nucleosomes. If one assumes that the DNA is uniformly wrapped around the nucleosomes, then by measuring the centre to centre spacing of the beads or the number of beads per micrometre it is possible to estimate the number of base pairs of B-conformation DNA that are packed into a micrometre length of the chromatin fibre. This is usually referred to as the 'packing ratio' (PR) of B-DNA to chromatin.

PR is defined as *the number of base pairs in 1 μm of chromatin divided by the number of base pairs in 1 μm of B-form DNA*. The number of base pairs in 1 μm of chromatin is then defined as *the number of nucleosomes per micrometre multiplied by the number of base pairs per nucleosome*. In the B-form of DNA the base pairs are 0.34 nm apart. There are, therefore, 2941 base pairs per micrometre of DNA.

In a transcription unit where the RNP fibrils are maximally packed and have a clearly defined gradient from short to long it is quite simple to measure the length of the transcription unit as the distance between the first and last definable transcript. It is possible to estimate the number of base pairs of DNA making up a transcription unit provided that the RNP transcripts and their associated polymerases are closely packed, since it is known that in such circumstances the arrangement of the DNA approximates to the B conformation. In *Xenopus laevis*, for example, the size of a 28 s + 5.8 s + 18 s ribosomal transcription unit is 7875 base pairs. As there are 2941 base pairs per micrometre of B-form DNA, the unit should be 2.67 μm long. The average measured length for ribosomal transcription units in *Xenopus* is 2.3 μm – a very good fit.

Analysis of transcription units is more difficult when the RNA polymerases are more widely and irregularly spaced. In such a situation the chromatin axis adopts a beaded nucleosomal arrangement between polymerases and when estimating the length of DNA in the transcription unit account must be taken of the packing ratio of DNA in nucleosomal regions between polymerases.

Laird *et al.* (1976) introduced a linear regression method for analysing irregular transcription units. The method involves measuring the lengths of RNP fibrils and plotting fibril length (*y* axis) against the distance of each fibril from a fixed point (*x* axis), normally the position of the first measurable transcript in the array. A plot of this kind is shown in Figure 7.11, constructed from data of Puvion-Dutilleul *et al.* (1978). A correlation coefficient can then be calculated. A positive and significant correlation coefficient will indicate that there is a relationship between the length of the RNP fibrils and their positions in the array with respect to a single point. An array of this kind can legitimately be called a transcription unit. More information about the transcription unit can be obtained by calculating the regression line for the data using the sum of least squares method. Then by applying the formula for a straight line, i.e. $y = a + b + x$, where a is the intercept on the y axis, b is the slope of the line, and x is the intercept on the x axis, the intercept on the y axis can be calculated, since when $y = O$, $x = -a/b$; x then marks the starting point of the transcription unit and the total length of the unit can be estimated by adding x to the distance between the first and the last transcript in the array.

In many cases an irregular array will not produce a significant correlation coefficient, or it will produce one that is low. One explanation for this might be that RNP fibrils originate from more than one point on the stretch of

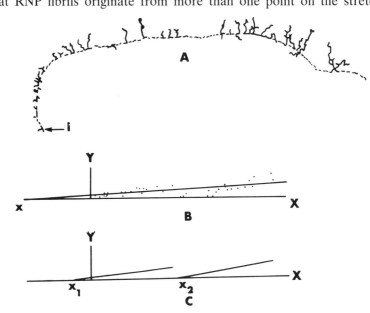

Figure 7.11 (A) A diagram of a non-ribosomal RNP fibril array from a liver cell, taken from the study of Puvion-Dutilleul *et al.* (1978); I = the first clearly definable transcript. (B) A graph of the RNP fibril length (y axis) against the distance of each fibril from I (x axis). From this plot a regression line and correlation coefficient can be calculated to determine the relationship between the two variables. In addition, x, the intercept of the regression line on the x axis, can be determined and used to estimate the overall length of the array. A correlation coefficient of 0.62 was calculated for the regression line in Figure 7.11(B). However, there is a stronger correlation between the lengths of the RNP fibrils and their position along the chromatin axis if two points of origin rather than one are assumed (x_1 and x_2) (Figure 7.11(C)). The correlation coefficient for the left-hand regression line is 0.87 and that for the right-hand one is 0.83. Regression analysis therefore indicates that the RNP fibrils in Figure 7.11(A) originate from two points and so may be taken to represent the products of two adjacent but separate transcription units

chromatin axis being examined, in which case more than one regression line can be drawn with different points of origin and perhaps with different slopes (Figure 7.11). Alternatively, a sudden change in the slope of the gradient of the RNP fibrils may occur due to processing of the RNP or changes in its secondary structure.

The slope of the regression line for most transcription units is less than 1, confirming that the length of an RNP fibril is not equivalent to the length of DNA from which it has been transcribed. RNA is complexed and packaged with protein as it is transcribed. The result is a beaded fibril appearing in Miller spreads with a thickness of 20–30 nm. The packing ratio of RNA to RNP varies for different transcription units within the same cell type and between different cell types. Beyer *et al.* (1981) have shown that specific packaging morphologies can be discerned in individual RNP fibrils and are specific for that transcription unit alone. Similarly, Laird *et al.* (1976) have described a distinctive plateau in the gradient of all ribosomal RNA gradients in *Drosophila* that may be related to the secondary structures found in the 28 s RNA part of the precursor.

VII. Concluding remarks

We have attempted to describe in considerable detail 'how to make a Miller spread' from a wide variety of cell types. We have offered a number of suggestions to aid the investigator both in the initial stages of handling the amphibian oocyte material and for later help in adapting the techniques to other cell types. The technique has, to date, been very successful in assaying the transcriptional activity of ribosomal RNA genes. 'All the rest' has been called non-ribosomal transcription since the particular genes one sees in whole nuclear preparations remain unidentified. Recently, Hutchison and Hamkalo (1981) and Hutchison *et al.* (1982) have adapted the *in situ* hybridization techniques for use in EM preparations. Thus, as gene-specific probes are developed, one may be able to answer questions about single copy genes, e.g. regarding their transcription as part of a larger heterogeneous RNA molecule, whether it is processed during transcription or following release from the DNA template, etc.

Other investigators have pioneered the localization of antibodies against chromosomal proteins in Miller spreads (e.g. McKnight *et al.*, 1977). Some laboratories are now trying these methods to view hormone binding sites adjacent to specific inducible genes.

The basic technique of 'Miller spreading' requires more time and less material than many biochemical methods of studying transcription. However, each kind of assay may be used to answer a certain kind of question. Miller spreads permit one to study RNA molecules as they are being transcribed, i.e. while they are still attached to the DNA template, and thus permit one to estimate the activity of single copies of a gene. Potentially one may also be able to correlate secondary and tertiary structures of RNP fibrils with specific proteins if, for example, one injects a particular gene into an oocyte or into tissue culture cells, then isolates the small transcribed injected molecules and analyses their RNP fibrils and attached proteins.

References

Bakken, A. H. (1976). Lampbrush chromosomes in mouse oocytes. *J. Cell Biol.*, **70**, 145a.

Bakken, A. H., and Hamkalo, B. A. (1978). Techniques for visualization of genetic material. In *Principles and Techniques of Electron Microscopy*, Vol. 9 (M. A. Hayat, ed.), Van Nostrand Reinhold Co., New York, pp. 84–106.

Bakken, A. H., Morgan, G., Sollner-Webb, B., Roan, J., Busby, S., and Reeder, R. H. (1982a). Mapping of transcription initiation and termination signals on *Xenopus laevis* ribosomal DNA. *Proc. Natl Acad. Sci.*, **79**, 56–60.

Bakken, A. H., Morgan, G., Sollner-Webb, B., Roan, J., Busby, S., and Reeder, R. H. (1982b). Mapping the regulators of *Xenopus laevis* ribosomal gene activity. In *Perspectives in Differentiation and Hypertrophy* (W. Anderson and W. Sadler, eds), Elsevier, New York, pp. 159–179.

Beyer, A. L., Bouton, A. H., and Miller, O. L., Jr (1981). Correlation of hnRNP structure and nascent transcript cleavage. *Cell*, **26**, 155–165.

Busby, S., and Bakken, A. H. (1979). A quantitative electron microscopic analysis of transcription in sea urchin embryos. *Chromosoma (Berl.)*, **71**, 249–262.

Busby, S., and Bakken, A. H. (1980). Transcription in developing sea urchins (*Strongylocentrotus purpuratus*): electron microscope analysis of cleavage, gastrula and prism stages. *Chromosoma (Berl.)*, **79**, 85–104.

Chooi, W. Y. (1979). The occurrence of long transcription units among the X and Y ribosomal genes of *Drosophila melanogaster*. *Chromosoma (Berl.)*, **74**, 57–74.

Foe, V. E., Wilkinson, L. E., and Laird, C. D. (1976). Comparative organization of active transcription units in *Oncopeltus fasciatus*. *Cell*, **9**, 131–146.

Francke, C., Edström, J.-E., McDowall, A. W., and Miller, O. L., Jr (1982). Electron microscopic visualization of a discrete class of giant translation units in salivary gland cells of *Chironomus tentans*. *EMBO J.*, **1**, 59–62.

Gall, J. G., Karrer, K., and Yao, M.-C. (1977). The ribosomal DNA of *Tetrahymena*. In *The Molecular Biology of the Mammalian Genetic Apparatus* (P. Ts'o, ed.), Elsevier/North-Holland Biomedical Press, Amsterdam, pp. 79–85.

Georgiev, G. P., Nedospasov, S. A., and Bakayev, V. V. (1978). Supranucleosomal levels of chromatin organization. In *The Cell Nucleus*, Vol. 6C, *Chromatin*, (H. Busch, ed.), Academic Press, New York and London, pp. 4–34.

Glatzer, K. H. (1975) Visualization of gene transcription in spermatocytes of *Drosophila hydei*. *Chromosoma (Berl.)*, **53**, 371–379.

Glatzer, K. H. (1979). Lengths of transcribed rDNA repeating units in spermatocytes of *Drosophila hydei*: only genes without an intervening sequence are expressed. *Chromosoma (Berl.)*, **75**, 161–175.

Glatzer, K. H. (1980). Regular substructures within homologous transcripts of spread *Drosophila* germ cells. *Exp. Cell Res.*, **125**, 519–523.

Gurdon, J. B. (1974). *The Control of Gene Expression*, Harvard University Press, Cambridge, Mass.

Hamkalo, B. A., and Miller, O. L., Jr (1973). Electron microscopy of genetic activity. *Annu. Rev. Biochem.*, **42**, 379–396.

Harper, F., and Puvion-Dutilleul, F. (1979). Non-nucleolar transcription complexes of rat liver as revealed by spreading isolated nuclei. *J. Cell Sci.*, **40**, 181–192.

Hill, R. S. (1979). A quantitative electron-microscope analysis of chromatin from *Xenopus laevis* lampbrush chromosomes. *J. Cell Sci.*, **40**, 145–169.

Hill, R. S., and Macgregor, H. C. (1980). The development of lampbrush chromosome-type transcription in early diplotene oocytes of *Xenopus laevis*: an electron microscope analysis. *J. Cell Sci.*, **44**, 87–101.

Hutchison, N. J., and Hamkalo, B. A. (1980). *In situ* hybridization at the electron microscope level. *J. Cell Biol.*, **87**, 45a.

Hutchison, N. J., Langer, P. R., Ward, D. C., and Hamkalo, B. A. (1982). *In situ* hybridization at the electron microscope level – Hybrid detection by autoradiography and colloidal gold. *J. Cell. Biol.*, **95**, 609–618.

Kressman, A., and Birnstiel, M. L. (1980). Surrogate genetics in the frog oocyte. In NATO Advanced Study Institute Series. Series A: Life Sciences, Vol. 31, *Transfer of Cell Constituents into Eukaryotic Cells* (J. E. Celis, A. Graessman, and A. Loyter, eds), Plenum, New York, pp. 383–407.

Labhart, P., and Koller, T. (1982). Structure of the active nucleolar chromatin of *Xenopus laevis* oocytes. *Cell*, **28**, 279–292.

Laird, C. D., and Chooi, W. Y. (1976). Morphology of transcription units in *Drosophila melanogaster*. *Chromosoma (Berl.)*, **58**, 193–218.

Laird, C. D., Wilkinson, L. E., Foe, V. E., and Chooi, W. Y. (1976). Analysis of chromatin-associated fiber arrays. *Chromosoma (Berl.)*, **58**, 169–192.

Lamb, M. M., and Daneholt, B. (1979). Characterization of active transcription units in Balbiani rings of *Chironomus tentans*. *Cell*, **17**, 835–848.

Macgregor, H. C. (1972). The nucleolus and its genes in amphibian oogenesis. *Biol. Rev.*, **47**, 177–210.

McKnight, S. L., and Miller, O. L., Jr (1976). Ultrastructural patterns of RNA synthesis during early embryogenesis of *Drosophila melanogaster*. *Cell*, **8**, 305–319.

McKnight, S. L., and Miller, O. L., Jr (1979). Post-replicative non-ribosomal transcription units in *D. melanogaster* embryos. *Cell*, **17**, 551–563.

McKnight, S. L., Bustin, M., and Miller, O. L., Jr (1977). Electron microscopic analysis of chromosome metabolism in the *Drosophila melanogaster* embryo. *Cold Spring Harbor Symp. Quant. Biol.*, **42**, 741–754.

Miller, O. L., Jr, and Bakken, A. H. (1972). Morphological studies of transcription. *Acta Endocrinol.*, Suppl. 168, 155–177.

Miller, O. L., Jr, and Beatty, B. R. (1969). Visualization of nucleolar genes. *Science*, **164**, 955–957.

Miller, O. L. Jr, Hamkalo, B. A., and Thomas, C. A. (1970). Visualization of bacterial genes in action. *Science*, **169**, 392–395.

Olins, A. L. (1977). *v* bodies are closely-packed in chromatin fibers. *Cold Spring Harbor Symp. Quant. Biol.*, **42**, 325–329.

Olins, A. L., and Olins, D. E. (1973). Spheroid chromatin subunits (*v* bodies). *J. Cell Biol.*, **59**, 252a.

Olins, A. L., and Olins, D. E. (1974). Spheroid chromatin units (*v* bodies). *Science*, **183**, 330–332.

Pruitt, S. C., and Grainger, R. M. (1981). A mosaicism in the higher order and structure of *Xenopus* oocyte nucleolar chromatin prior to and during ribosomal gene transcription. *Cell*, **23**, 711–720.

Puvion-Dutilleul, F., Bernhard, A., Puvion, E., and Bernhard, W. (1977). Visualization of two different types of nuclear transcriptional complexes in rat liver cells. *J. Ultrastruct. Res.*, **58**, 108–117.

Puvion-Dutilleul, F., Puvion, E., and Bernhard, W. (1978). Visualization of non-ribosomal transcription complexes after cortisol stimulation of isolated rat liver cells. *J. Ultrastruct. Res.*, **63**, 118–131.

Rattner, J. B., and Hamkalo, B. A. (1978a). Higher order structure in metaphase chromosomes: I. The 250 Å fiber. *Chromosoma (Berl.)*, **69**, 363–372.

Rattner, J. B., and Hamkalo, B. A. (1978b). Higher order structure in metaphase chromosomes: II. The relationship between the 250 Å fiber, superbeads and beads-on-a-string. *Chromosoma (Berl.)*, **69**, 373–380.

Stone, G. E. and Cameron, I. L. (1964). Methods for using *Tetrahymena* in studies of the normal cell cycle. In *Methods in Cell Physiology*, Vol. 1 (D. M. Prescott, ed.), Academic Press, New York, pp. 127–140.

Thoma, F., and Koller, Th. (1981). Unravelled nucleosomes, nucleosome beads and higher order structures of chromatin: influence of nonhistone components and histone H1. *J. Mol. Biol.*, **149**, 709–734.

Thoma, F., Koller, Th., and Klug, A. (1979). Involvement of histone H1 in the organization of the nucleosome and of the salt-dependent superstructures of chromatin. *J. Cell Biol.*, **83**, 403–427.

Trendelenburg, M. F., and Gurdon, J. B. (1978). Transcription of cloned *Xenopus* ribosomal genes visualized after injection into oocyte nuclei. *Nature (Lond.)*, **276**, 292–294.

Trendelenburg, M. F., Mathis, D., and Oudet, P. (1980). Transcription units of chicken ovalbumin genes observed after injection of cloned complete genes into *Xenopus laevis* oocyte nuclei. *Proc. Natl Acad. Sci.*, **77**, 5984–5988.

Whiteley, A. H., and Mizuno, S. (1981). Electron microscope visualization of giant polysomes in sea urchin. *Wilhelm Roux's Arch. Develop. Biol.*, **190**, 73–82.

Wickens, M. P., and Laskey, R. A. (1981). Expression of cloned genes in cell-free systems and in microinjected *Xenopus* oocytes. *Genet. Eng. Ser.*, **1**, 103–167.

Wickens, M. P., Woo, S., O'Malley, B. W., and Gurdon, J. B. (1980). Expression of a chicken chromosomal ovalbumin gene injected into frog oocyte nuclei. *Nature (Lond.)*, **285**, 628–634.

Woodcock, C. L. F. (1973). Ultrastructure of inactive chromatin. *J. Cell Biol.*, **59**, 368a.

Worcel, A. (1977). Molecular architecture of the chromatin fiber. *Cold Spring Harbor Symp. Quant. Biol.*, **42**, 313–324.

8

In situ hybridization

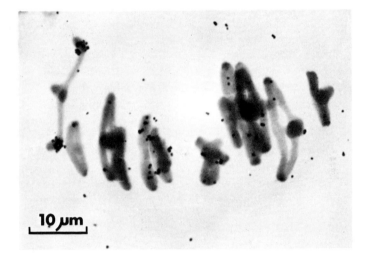

10 μm

The photograph shows an autoradiograph of a squash preparation of a
set of first meiotic metaphase chromosomes from the Red-backed salamander, *Plethodon cinereus*, after *in situ* hybridization with nick-translated ³H-labelled ribosomal DNA. The clusters of silver grains mark the
positions where the labelled rDNA has bound to complementary sequences on the chromosome, marking the sites of the main nucleolus
organizers in this animal. [Reproduced with permission from Macgregor,
H. *et al.*, *Chromosoma (Berl.)*, **59**, 283–299 (1977)]

The objective of the *in situ* nucleic acid hybridization technique is the detection or localization of specific DNA or RNA sequences in cells, cell nuclei,
or chromosomes employing a radioactive single-stranded nucleic acid 'probe'
of complementary base sequence, and the standard technique of autoradiography. In the simplest terms a normal microscope preparation is made em-

ploying sectioning, squashing, or air-drying techniques. If RNA is to be detected the preparation is immediately incubated in a solution of single-stranded radioactive probe, washed, and autoradiographed. If DNA is to be detected then the preparation is treated for removal of RNA, then treated to denature the chromosomal DNA, and then incubated in labelled single-stranded probe, washed, and autoradiographed.

The method can be employed for quite precise localization of gene sequences in chromosomes, experiments having been directed at the localization of genes for ribosomal RNA, 5sRNA, transfer RNA, and histones, as well as a wide variety of repetitive and single-copy DNA sequences. The most successful results have been obtained with sequences that are repeated many times over in one or a few places in the chromosome set, thus providing a relatively large binding site or 'target' for the probe molecules. However, the detectability of hybrid molecules in cytological preparations depends largely on the level of radioactivity of the probe and on the efficiency of the hybridization reaction and the autoradiographic procedure. Recent developments in all these areas have brought the detection of single copy or low copy number genes by *in situ* hybridization well within reach and several recent studies have shown this to be entirely possible. Of course it has always been possible to probe for and detect low copy number genes in giant polytene chromosomes, since these provide a ready made situation in which all gene sequences are multiplied several hundred times over, all copies remaining localized and concentrated in their rightful place on the chromosome set. In fact, DNA to DNA *in situ* hybridization can be employed with confidence for the localization of any moderately long sequences that are repeated more than 100 times over at one place in a chromosome set. It can also be used for detection of low copy number genes, but success depends on having a hot and pure probe, very good and critical technique, and generally using every available means to reduce background and encourage specific annealing between probe and target sequences, and enhance the sensitivity of detection.

In the early days of *in situ* hybridization it was common to employ *in vivo*-labelled ribosomal RNA for detection of nucleolus organizers, or complementary RNA (cRNA) for detection of other highly repetitive sequences. *In vitro* labelling with iodine-125 was also employed but not generally accepted as a useful method. The cRNA was produced by copying a DNA template with RNA polymerase from *Esherichia coli*. Specific activities ranged from $\sim 10^5$ counts min^{-1} μg^{-1} for material labelled *in vivo* to $\sim 10^7$ counts min^{-1} μg^{-1} for cRNA. Always the objective was to locate chromosomal or nuclear DNA sequences. A little later, beginning with some experiments by Pukkila (1975), DNA probes began to be used for the detection of intracellular RNA, especially that which was still associated with the gene from which it had just been transcribed. This called for a method of labelling

DNA to specific activities that were useful for *in situ* hybridization. Nick translation of DNA with DNase I and DNA polymerase I quickly became fashionable and has since proved to be the most successful and generally useful method for obtaining good hot tritium-labelled probes.

The cytological preparations used for *in situ* hybridization can be made in any way that leaves the target DNA or RNA sequences hybridizable. In most cases, experiments are designed around the use of 3 : 1 fixed material that has either been squashed in 45% acetic acid and air dried from 95% ethanol or splashed on to slides and dried directly from the fixative. The technique has also been widely used on lampbrush chromosome preparations that have been made in the way described in Chapter 4 and then fixed in ethanol or ethanol–acetic acid and air dried. There is no reason why *in situ* hybridization should not be applied to sectioned material. However, there is one general rule. The material on the slide should never have been exposed to any fixative that is likely to interfere with the hybridization reaction, either by preventing denaturation of the DNA prior to hybridization or by radically altering and reinforcing the binding of DNA to protein. Preparations are normally made on subbed slides in consideration of subsequent autoradiography, but this may not be entirely necessary. The general experience is that subbing does not affect the hybridization reaction in any way and does eliminate the risk, however marginal, of losing preparations through slipping of autoradiographic emulsion. In one series of experiments (Gerhardt *et al.*, 1981) slides were treated with 10× Denhardt's solution in an attempt to reduce absorption of labelled probe to the glass. Denhardt's solution (×10) contains 0.2% bovine serum albumin, and this will doubtless have the same effect as the gelatin in subbing mixture.

In what follows, three protocols will be given. The first is specifically recommended for the localization of nucleolus organizers on chromosomes using nick-translated ribosomal DNA (rDNA) – considered to be a good example of DNA to DNA hybridization. The second is recommended for localization of RNA transcripts that are still attached to their chromosomal DNA template, again using a nick-translated DNA probe. The third covers the use of ^{3}H-labelled complementary RNA (cRNA) as a probe for localization of chromosomal DNA sequences.

A fourth approach involves the use of labelled cRNA probes synthesized *in vitro* from single-stranded DNAs such as can be produced by cloning DNA sequences into M13 phage. The method is clearly of immense experimental value, but the production of the template for the probe is quite complicated and beyond the scope of this book. A good example of the application of the method can be found in Diaz *et al.* (1981).

After the three protocols there will be a general discussion of each step in the technique together with notes on some of the useful modifications that have been introduced from time to time. Finally, some details will be given of modifications and additions to the basic technique that are essential if *in situ* hybridization is to be used for the detection of single copy sequences.

I.   DNA to DNA, repeated sequences

The following equipment and materials are needed:

(a) Moist chambers (a square plastic Petri dish with a filter paper in the bottom thoroughly wetted with an excess of the same solution as that with which the preparations are to be in contact, and a piece of glass rod bent into the shape of a U to act as a support to keep the slides from coming in contact with the wet filter paper)

(b) Micropipetting system, 10–200 μl

(c) Coverglasses 18 or 22 mm square, number 1½

(d) pH papers

(e) Incubator, 30–70 °C

(f) 20× SSC (SSC = 0.15 M NaCl, 0.015 M trisodium citrate, pH 7.0)

(g) 70% ethanol

(h) 95% ethanol

(i) Formamide: Deionize the formamide by adding 2–3g mixed bed resin (AG 501–X8(D), 20–50 mesh, Bio-Rad Laboratories), stirring for 30 min and filtering through Whatman No. 1.

(j) 0.07 M NaOH

(k) 0.1 M NaOH

(l) 0.1 M HCl

(m) 5% trichloroacetic acid (TCA)

(n) Ribonuclease A (Sigma cat. no. R4875), 1 mg/ml stock

(o) Ribonuclease T1 (Sigma cat. no. R8251), 1000 units/ml stock

The ribonuclease A stock should be made up in 0.02 M sodium acetate, pH 5, and heated to near boiling for 10 min by placing in a boiling water bath. This step inactivates contaminating DNase. The stock solution can then be stored frozen and diluted to appropriate strength when needed.

A.   Preparation of probe

At the start of this procedure the probe DNA should be freshly precipitated from ethanol and vacuum dried, or it should be freeze dried. It should be in the bottom of a small glass or plastic vial (Eppendorf vials are entirely suitable). Parallel-sided glass vials measuring about 5 cm × 1 cm are suitable. Ascertain the number of slides that are to be processed. If 22 mm square coverglasses are to be used then each slide will require 30 μl of reaction mix. If 18 mm square or smaller coverglasses can be used then each preparation will require correspondingly less reaction mix. As a rule 18 mm square coverglasses require 20 μl of mix, and 1 cm square coverglasses require 15 μl. If it is necessary to use smaller amounts of reaction mix for each preparation, then special precautions must be taken to prevent evaporation during incubation. Some investigators use as little as 5–10 μl of reaction mix for 18 or 22 mm square coverglasses and avoid evaporation by sealing round the edge of the coverglass with rubber cement, nail varnish, or a 1 : 1 mixture of paraffin and vaseline. Such small amounts of reaction mix are successful when hybrids are incubated at 37 °C, but definitely will not work with incubations at 65 °C.

The reaction mix is then made up according to the amount that will be needed for all the slides that are to be processed.

In what follows, assume 30 μl mix per preparation, 100 000 counts of probe per preparation, and a total of 10 preparations to be processed.

(1) Start by carefully titrating the 0.1 M NaOH against the 0.1 M HCl to determine *exactly* how much HCl is needed precisely to neutralize a given amount of NaOH. For the present it will be assumed that the two solutions are exactly equivalent, but this is rarely the case in practice.

(2) Dissolve enough DNA probe to provide a total of 3×10^6 counts/min (as determined by liquid scintillation counting) in 30 μl of 0.1 M NaOH. Take special care to ensure that all the DNA is completely dissolved. This is best accomplished by rolling the liquid round and round the bottom of the vial or by flicking the vial for several minutes at room temperature.

(3) Add 150 μl of formamide (95+% stock) and mix well: final concentration will be 50%.

(4) Add 60 μl of 20× SSC and mix well: final concentration will be 4× SSC.

(5) Add 30 μl of water and mix well.

(6) Cool on ice for several minutes.

(7) Add 30 μl of 0.1 M HCl and mix well.

(8) Keep ice cold and use to set up *in situ* hybrids within 10–15 min. Speed is of some importance at this stage, especially when working with highly repeated DNA probes that may commence reassociation before the *in situ* hybrid is set up.

The most satisfactory arrangement is to prepare the probe during the final stages of preparation of the slides, immediately after they have been denatured and during the subsequent washing and drying procedures.

B. Preparation of the slides

Start with preparations that have been air dried on subbed slides from 95% ethanol or from 3 : 1 ethanol (or methanol)–acetic acid.

(1) Endogenous RNA will act as a competitor to the hybridizing DNA and by binding some of the labelled probe it will also tend to increase unwanted autoradiographic background labelling. It must therefore be removed by treating with ribonuclease. Place the slides for 2 h at 37 °C in a solution containing 50 μg/ml of RNase A and 100 units/ml of RNase T1 in 2× SSC. If it is important to economize on enzyme then the digestion can be carried out by placing about 30 μl of enzyme solution on the preparation, covering with a coverglass, and placing the slide in a moist chamber in an oven at 37 °C.

(2) Wash three times in 2× SSC; each wash should last 10 min.

(3) Place the slides in 0.07 M NaOH at 20 °C for 3 min.

(4) Wash three times in 70% ethanol.

(5) Wash twice in 95% ethanol and air dry.

The slides are now ready for the reaction mix and hybridization.

Denaturation in NaOH and final washes should be carried out immediately before setting up the hybridization reaction. It is generally convenient to prepare the probe while the slides are washing and drying after treatment with NaOH.

C. Hybridization

Immediately before putting the hybridization mix on the slides, check its pH with pH paper. It should be 7.0 ±0.2. If it is not, then something may have gone wrong during the preparation of the probe and it is best to stop at this point and check before committing further time and effort.

Likewise, check the number of counts in the final reaction mix. This can be done by taking 5 or 10 μl of mix, diluting it tenfold with water, and then putting 10 μl of that on a Millipore filter and counting the filter in a liquid scintillation system. Dilution of the mix is necessary because the high formamide and salt concentrations affect both the filter and the counting efficiency. If the radioactivity of the probe (in counts per minute) is not what was expected then stop and check before setting up the hybridization.

(1) Set out the slides in the moist chambers, making quite sure that each has the preparation uppermost.

(2) Very carefully clean and set out as many coverglasses as there are slides.

(3) Using a micropipette place 30 μl of reaction mix in the form of a single drop right in the middle of the preparation area of each slide. If there should be any difficulty in seeing the area of the preparation then it may help to breathe – 'huff' – lightly on the slide. The preparation will show up as the area that mists up more thickly than other parts of the slide.

(4) Place a coverglass on each preparation, making sure that it is positioned to cover the whole preparation.

(5) Put the tops on the moist chambers and place the chambers in the 37 °C incubator for 6–12 h.

D. Washing after hybridization

(1) Pick up each slide and dip it into a large beaker filled with 2× SSC. The coverglass will fall off.

(2) Place the slides together in a rack in fresh 2× SSC.

(3) Place the slides in 2× SSC at 65 °C for 15 min.

(4) Wash twice in 2× SSC at room temperature, each wash of 10 min.

(5) Place in 5% TCA at 5 °C for 5 min.

(6) Wash twice in 2× SSC at room temperature, each wash lasting 10 min.

(7) Wash twice in 70% ethanol, each wash lasting 10 min.

(8) Wash twice in 95% ethanol, each wash lasting 10 min.

(9) Air dry.

The slides are now ready for autoradiography.

Notice that this is the simplest possible procedure for DNA to DNA hybridization. There are other ways of denaturing both probe and preparation. The reaction mix and washing procedures can be varied according to the stringency of hybridization that is required. Various additional steps and reagents can be introduced to reduce background or non-specific binding. Some of these matters are discussed below. However, the procedure outlined above has given excellent results in the authors' hands over a period of years and in experiments with a wide variety of probes and cytological specimens.

II. DNA (probe) to RNA (chromosomal), repeated sequences

A. Preparation of probe

Proceed exactly as described in the protocol for DNA to DNA hybridization.

B. Preparation of the slides

Do nothing. The slides are ready for hybridization as soon as they have been air dried from ethanol or ethanol–acetic acid.

C. Hybridization

As in DNA to DNA protocol.

D. Washing after hybridization

As in DNA to DNA protocol.

In essence, the only differences between DNA to DNA and DNA to RNA hybridization are that the slides are not treated with ribonuclease and not denatured, for obvious reasons.

III. RNA (probe) to DNA (chromosomal), repeated sequences

A. Preparation of the probe

The RNA solution can be made up in 2× SSC or in 4× SSC–50% formamide at a concentration to provide 50 000–100 000 counts/min for each 30 μl of final reaction mix.

B. Preparation of the slides

Exactly as in DNA to DNA protocol.

C. Hybridization

As in DNA to DNA protocol: at 65 °C for 6–12 h if the probe is in 2× SSC, and at 37 °C for 6–12 h if the probe is in 4× SSC–50% formamide.

D. Washing after hybridization

(1) Pick up each slide and dip it into a large beaker filled with 2× SSC. The coverglass will fall off.
(2) Place the slides together in a rack in fresh 2× SSC. Leave for 15 min at room temperature.
(3) Transfer to ribonuclease (50 μg/ml RNase A, 100 units/ml RNase T1 in 2× SSC) at 37 °C for 1 h.
(4) Wash twice in 2× SSC, 15 min each wash.
(5) Place in 5% TCA at 5°C for 5 min.
(6) Wash twice in 2× SSC, 10 min each wash.
(7) Wash twice in 70% ethanol, 10 min each wash.
(8) Wash twice in 95% ethanol, 10 min each wash.
(9) Air dry.

The slides are now ready for autoradiography.
Important: These protocols should be carried out without interruption. On no account should the procedure be stopped after either the probe or the

preparations have been denatured. A useful work schedule involves commencing the denaturation of the probe and preparations at around 10 p.m. and setting up the hybridization to run overnight, with washing starting at about 9 a.m. the following morning.

IV. Comments on methods

Good *in situ* hybrids begin with good cytological preparations. Squashes and splashes should be made such that nuclei and chromosomes are well separated from one another and surrounded by the minimum possible amount of cytoplasm and tissue. In places where there is a high density of material and considerable thickness of tissue on the slide, the hybridization reaction will be inefficient and there will be a high autoradiographic background due to non-specific binding of the radioactive probe.

Choice of slides and coverglasses may be important. Some slides and coverglasses have a high soda content. The hybridization reaction takes place in a small liquid volume between slide and coverglass, and the alkaline nature of the glass may raise the pH of the reaction mix to a point at which hybrid molecules are no longer stable. Slides supplied by Arthur Thomas and Corning are known to be satisfactory. With slides and coverglasses from other sources it may be advisable to check their effect on pH under hybridization conditions. If in doubt then cook them in 1 M HCl at 60 °C for 6–12 h, then wash thoroughly in water and oven dry them at over 80 °C.

There is considerable controversy over the question of storage of slides before *in situ* hybridization. Some investigators claim that storage makes no difference so long as preparations are kept dry. Others advise storage in the cold and under nitrogen. Still others say slides should not be stored longer than is absolutely necessary. There is no doubt that cytological preparations do change in the course of dry storage. Those who use Giemsa banding techniques call the change 'ageing'. Of course it is not known whether this kind of change affects hybridizability, but in any event it would seem to be sensible to make a rule of setting up *in situ* hybrids as soon after making the preparations as is reasonably possible. If the slides must be stored, then there would seem to be no harm in storing them under conditions that are least likely to cause deterioration. Dryness and low temperature are most important in this regard.

Some investigators have advocated the wearing of gloves for handling all preparations to be *in situ* hybridized, mainly to avoid contamination with 'finger-borne ribonuclease'. The risk of such contamination is probably much less significant than the risk of radioactive contamination from handling the hot probes. So gloves would seem to be a sensible precaution all round.

Many variations of the pretreatment of slides have been suggested and tried. Ribonuclease concentrations vary from 10 to 100 µg/ml. The use of T1

RNase is not common practice and is suggested here merely because it has proved effective and should give more complete removal of RNA than RNase A alone. Some persons have recently introduced an acetylation step before hybridization (Hayashi *et al.*, 1981) with the claim that this significantly reduces background due to non-specific binding. Acetylation is carried out as follows:

(1) Place the slides in 0.1 M triethanolamine (adjusted to pH 8 with 0.1 M HCl).

(2) Stir vigorously and at the same time add acetic anhydride to a concentration of 0.5%.

(3) Stop the stirring and leave for 10 min at room temperature.

(4) Wash twice in 2× SSC, each wash to last 10 min.

(5) Wash twice in 70% ethanol, each wash to last 10 min.

(6) Wash twice in 95% ethanol, each wash to last 10 min.

(7) Air dry.

Pretreatment of the slides at 70 °C for 30 min in 2× SSC with subsequent washing in 70% and 95% ethanol and air drying is said to improve the attachment of nuclei and chromosomes to the slide and ensure against loss of material during processing. However, the usefulness of this step is questionable. Another step that has been said to be almost essential if material is not to be washed from the slide in subsequent processing is the baking of dried preparations in an oven at 65 °C before hybridization. This step is certainly not necessary and it is now known that those who advocated it were omitting to bake their slides after subbing and before making their squash preparations. The baking of preparations was therefore merely a delayed baking of the subbing mixture. In fact, nothing need be lost from a preparation during *in situ* hybridization provided that the slide has been properly subbed and the squash or splash properly made.

Denaturation of the cytological preparations can be carried out in several ways. The use of 0.07 M NaOH for 3 min at room temperature is simple and effective but not conducive to good cytological preservation. The same may be said of heat denaturation. Investigators must simply try for themselves and see which gives the best results with their particular material. Denaturation with 0.1 M HCl at 20 °C for 10 min gives excellent cytological preservation and adequate denaturation, although the level of hybridization is certainly not as high as after heat or alkali denaturation. A good account and discussion of the effects of different denaturing agents is given by Singh *et al.* (1977).

The composition of the reaction mix and the temperature at which the hybridization is carried out determines the speed with which hybrid molecules form and the stringency of the reaction conditions with regard to mismatching between probe and target sequences. As a rule, the speed with which nucleic acid duplexes form is markedly dependent upon salt concentration and increases steeply up to about 1 M salt. Likewise the rate of the reaction rises with temperature until the melting temperature of the duplex is approached. The problem here is to select conditions that give optimum speed of reaction, minimum level of mismatching without discouraging hybrid formation altogether, and minimum cytological damage as a result of high salt or high temperature. The formula 2× SSC and 65 °C works well but does cause some cytological damage and even hydrolysis of nucleic acid. Formamide lowers the melting temperature of nucleic acids and its addition to the hybridization mix permits the use of much lower incubation temperatures without sacrificing stringency. Cytological damage by high temperature is therefore avoided.

The conditions under which *in situ* hybrids are washed affect the outcome in just the same way as do the conditions under which the hybridization reaction is carried out. Stringent washing that would leave only well matched hybrids intact would involve low salt concentrations and high temperatures. Poorly matched hybrids would be destroyed. However, the main purpose of washing is to remove probe molecules and their degradation products that are not bound to complementary sequences on nuclei or chromosomes. So on the one hand washing is needed to remove poorly matched hybrids, and on the other hand it is needed to get rid of unbound or non-specifically bound material. As a rule washing in 2× SSC at room temperature and at 65 °C accomplishes both these objectives, and the addition of a quick treatment with cold TCA helps to make doubly sure of the elimination of unbound low molecular weight material.

A few investigators favour the inclusion of Denhardt's solution in both the reaction mix and the washing procedure. Denhardt's solution consists of bovine serium albumin, Ficoll, and polyvinylpyrrolidone all at 0.02%, and was originally introduced in 1966 as a mixture that prevents the adsorption of single-stranded DNA to nitrocellulose filters. The benefits of including Denhardt's solution are likely to be marginal with *in situ* hybridization, and especially so when trying to locate repeated DNA sequences. Moreover, there is some risk of cytological damage if Denhardt's is used above 1× strength.

V.   Some comments about the reaction

Several authors have discussed the conditions of an *in situ* hybridization reaction with regard to matters like saturation and efficiency. Articles by Gall and Pardue (1971) and Wimber and Steffensen (1973) should be consulted in the first instance, and two papers by Szabo *et al.* (1975, 1977) are particularly noteworthy in the sense that they represent the only published accounts of actual experimental studies on the technique of *in situ* hybridization. These authors have systematically examined the reproducibility of the technique, the characteristics of the hybrid molecules formed during a reaction, the kinetics of the reaction, the efficiency of the reaction, several autoradiographic parameters, and the significance of a wide range of experimental variables.

There is not much that can usefully be added here on this topic. The facts are that *in situ* hybridization works, that it is entirely reproducible with the same material and the same probe, that it is generally about 6–10% efficient, and that the kinetics and efficiency of the reaction will differ according to the size and complexity of the sequences being hybridized. It is a costly operation in terms of both time and materials. A newcomer to the technique is perhaps best advised to follow one of the simpler protocols suggested in this chapter, with regard to whether repeated or single copy sequences are involved; but before designing definitive experiments all the main literature sources cited above should be carefully studied. It should then be possible to make useful calculations as to the likelihood of success and the quantities and timings that will be needed in the experiment.

VI.   DNA to DNA, single copy sequences

The detection of single copy DNA sequences on mitotic chromosomes requires some special methods. Three objectives must be met. First, probes must be pure, well characterized, and reliable. Secondly, special steps must be taken to bring probe and target sequences together and encourage fast annealing. Thirdly, non-specific binding of labelled probe to any part of the preparation must be reduced to an absolute minimum. This is of the greatest importance since the autoradiographic signal from the site of specific annealing will always be weak and evaluation of an experiment will usually rely heavily on statistics of grain counting and distribution. The first objective has been met by the use of recombinant DNA probes and now presents few problems. The second objective has been met by the use of dextran sulphate. The third objective has been tackled in a variety of ways at almost every stage in the procedure. Only the use of dextran sulphate and the prevention of non-specific binding will be dealt with here.

Dextran sulphate was introduced into nucleic acid hybridization some years ago as an agent that effectively increases the rate of annealing of DNA in solution by about 10-fold, especially when randomly cleaved (nick-translated) DNA is used. In theory and in fact, the 10-fold increase in reassociation rate is attributable to the exclusion of DNA from the volume of the solution occupied by the dextran sulphate, so that local concentrations of DNA at the molecular level are greatly increased. This not only promotes faster reassociation between probe and target but it also encourages the formation of extensive networks of DNA wherever, after denaturation, two partially complementary fragments can reanneal to form a partial duplex. So not only is reassociation between probe and target speeded up, but network formation greatly increases the amount of probe and associated DNA that will bind to the target and contribute to the autoradiographic signal. Discussions of the usefulness and effect of dextran sulphate in solution and in *in situ* hybridizations are given by Wahl *et al.* (1980) and Lederman *et al.* (1981) respectively. There is, of course, no reason why dextran sulphate should not be used for *in situ* hybrids with repeated sequences, but its principle value in recent times has been in extending the application of the *in situ* hybridization technique to the detection of single or low copy number genes.

With regard to reducing background or non-specific binding of probe, all manner of measures have been introduced. In the main, the attack has been concentrated on the binding of probe to the glass of microscope slides and the non-specific annealing of probe sequences to other DNAs in the target tissue. The first factor has largely been tackled through the use of Denhardt's solution and the inclusion of cold competitor DNA in the hybridization mix. Non-specific annealing has been discouraged by the use of denaturation, hybridization, and washing procedures that are much more complex and fastidious than those employed when hybridizing to more easily localized repetitive DNA sequences.

The following is a selection of procedures that have been adopted by persons who have successfully localized single copy genes by *in situ* hybridization of ^{3}H-labelled DNA to mitotic metaphase chromosomes:

(a) Pretreatment of microscope slides: 5 h at 60 °C in 10× Denhardt's solution. 1× Denhardt's solution contains 0.02% bovine serum albumin (fraction V, Sigma cat. no. A-4503), 0.02% Ficoll (Sigma cat. no. F-4375), 0.02% polyvinylpyrrolidone (Sigma cat. no. PVP-360).

(b) Denaturation of the chromosomal DNA: 70% formamide, 2× SSC, pH 7.0, 70 °C, 2 min.

(c) Hybridization mix: labelled DNA at not more than 50 μg/ml, 1000-fold excess of unlabelled denatured salmon sperm DNA (Sigma cat. no. D-1626), 50% formamide, 2× SSC, 10% dextran sulphate pH 7, 1× Denhardt's solution.

(d) Denaturation of the DNA probe: incubation of the hybridization mix at 70 °C for 15 min.

(e) Hybridization: 37 °C for 12 h.

(f) Post-hybridization washes: 50% formamide, 2× SSC, 39 °C, followed by room temperature washes in 2× SSC, ethanol, etc., and air drying. In some cases more stringent washes have been used, including 2× SSC at 65 °C for up to 30 min, and 0.1× SSC at 4 °C overnight.

Experiments in which these sorts of procedures have been applied have produced significantly positive autoradiographs, albeit sometimes with only a single silver grain over the target area, after autoradiographic exposure times of between 2 and 4 weeks.

Useful references on *in situ* hybridizations involving single copy genes are Harper and Saunders (1981), Harper *et al.* (1981), Lederman *et al.* (1981), Gerhardt *et al.* (1981), and Malcolm *et al.* (1981a, b).

VII.   Methods for labelling probes

A.   Complementary RNA (Gall and Pardue, 1971)

Make ready the following items:

(a) Dried 'XTPs', tritium-labelled with specific activities of 10–30 Ci/mM: 100 μCi GTP; 100 μCi ATP; 100 μCi CTP; 100 μCi UTP. Put all of these in a small glass tube, vacuum dry and store in the freezer.

(b) β-Mercaptoethanol, stock solution (Sigma cat. no. M-6250, usually about 14.3 M), diluted 20 μl in 10 ml water.

(c) Salts solution: 1 M Tris, pH 7.9, 4.0 ml; 2 M KCl, 9.4 ml; 1 M MgCl$_2$, 0.58 ml; 0.01 M EDTA, 0.88 ml. Make up to a final volume of 25 ml with water. Add 80 μl of 0.125 MnCl$_2$ per millilitre of salts solution just before use.

(d) Sephadex column, 15 cm × 1 cm bed volume, Sephadex G-50 made up in water, takes about 5 g dry weight of Sephadex. An alternative arrangement is to use Bio-gel P-60 (Biorad Laboratories cat. no. 150-1640) in a small plastic 'Quick-sep' column (Isolab Inc. cat. no. qs-p).

(e) Enzymes: *E. coli* RNA polymerase (Sigma cat. no. R5376); DNase I (Sigma cat. no. D-1126), 1 mg/ml in water

(f) Sarkosyl or sodium dodecyl sulphate (SDS) 5% in water

(g) *E. coli* tRNA, 50 μg/ml in 0.04 M Tris, pH 7.9 (Sigma cat. no. R1753)

(h) Phenol, water-saturated

The procedure is then as follows:

(1) Prepare DNA solution in 0.04 M Tris, pH 7.9, to have an optical density at 260 nm (OD$_{260}$) of between 1.0 and 2.0 (50–100 μg/ml). (Place DNA in a boiling water bath for 5 min and cool on ice if transcription from single-stranded DNA is required.)

(2) To the dried XTPs add the following in order: DNA, 100 μl; β-mercaptoethanol (dilute), 50 μl; salts solution 50 μl; water 40 μl; RNA polymerase 10 μl. The total volume should be 250 μl. The volume of the

DNA solution and the water can be varied, so long as the two combined add up to 140 μl and contain 5–10 μg of DNA.

(3) Cover the tube with Parafilm and incubate at 37 °C for 60–90 minutes.

(4) Add 0.75 ml of 0.04 M Tris, pH 7.9, and 20 μl of DNase I (= 20 μg). Hold at 20 °C for 15 min. A slight fibrous precipitate may form. Ignore it.

(5) Remove two 10 μl samples and spot on to nitrocellulose filters. Dry the filters. Then place one filter in cold 5% TCA for 5 min, wash it in 70% ethanol and redry it. Count both filters in a liquid scintillation system. Calculate the percentage of counts which are TCA insoluble; this is the efficiency of incorporation and should be between 15 and 40%.

(6) Add 200 μl of 5% Sarkosyl or SDS.

(7) Add 50 μg of *E. coli* RNA in 1 ml of 0.04 M Tris, pH 7.9. This is added as a 'carrier' in cases where cRNA is being transcribed from eukaryotic DNA.

(8) Add 2 ml of water-saturated phenol. The phenol should be redistilled. Alternatively use a mixture of 100 g of phenol, 0.1 g of 8-hydroxyquinoline, 100 ml of chloroform, and 4 ml of isoamyl alcohol.

(9) Centrifuge for 10 min at about 10 000 rev/min. Remove the supernatant, and re-extract the phenol with a little more 0.04 M Tris (about 200 μl).

(10) Lower the water level on the top of the Sephadex column to the top of the resin bed. Gently add the supernatant. Allow the water to flow through the column and collect 25 consecutive, numbered 30-drop fractions.

(11) Spot 10 μl from each fraction on to a nitrocellulose filter, dry the filters, and count.

(12) Pool the peak fractions. They are usually around numbers 13–18.

(13) Place the pooled fractions in an 85 °C water bath for 3 min.

(14) Force through an 0.45 μm pore nitrocellulose filter, estimating the total volume finally collected.

(15) Spot 10 μl of the final product on a nitrocellulose filter for measurement of the number of counts per minute per millilitre.

Note that if separation of labelled fractions is carried out with P-60 and a small Quik-sep column then there will be fewer and smaller fractions, but the procedure will be the same in principle.

Store at −20 °C.

The cRNA is now ready for use but may have to be concentrated by freeze drying.

B.   Nick-translation of DNA

This procedure (Maniatis *et al.*, 1975; Macgregor and Mizuno, 1976; Rigby *et al.*, 1977; Lai *et al.*, 1979) is specifically for 1 μg of DNA in 100 μl of reaction mix. It may be expected to yield DNA with specific activities of 2–6 $\times$ 10^6 counts min^{-1} μg^{-1} or 0.8 $\times$ 10^7 – 2.4 $\times$ 10^7 disint. min^{-1} μg^{-1}.

(1) Take 2 $\times$ 10^{-3} μM each of [^{3}H]dATP, [^{3}H]dCTP, [^{3}H]dGTP, and [^{3}H]TTP and mix them together in a small glass or plastic vial. An example of such a mixture is as follows:

	mCi/ml	mCi/μM	μM/ml	2×10^{-3} μM
d8 [^{3}H]ATP	1	17.6	0.057	0.035 ml
d8 [^{3}H]CTP	1	20.0	0.050	0.040 ml
d8 [^{3}H]GTP	1	21.0	0.047	0.042 ml
M– [^{3}H]TTP	1	30.0	0.033	0.060 ml

The total volume of this mixture is 0.177 ml.

(2) Put 0.018 ml of the mixture in a small vial and freeze dry completely. An Eppendorf vial is ideal for this operation.

(3) Make up fresh from stocks a salt solution containing the following: 2.5 ml of 1 M Tris-HCl, pH 7.9; 0.5 ml of 1 M mercaptoethanol; 0.25 ml of bovine serum albumin, 10 mg/ml; 1.5 ml of distilled water; 0.25 ml of 1 M MgCl$_2$ (add this last of all). The total volume is 5 ml.

(4) Prepare the DNA solution to have a concentration of between 200 and 500 μg/ml (OD$_{260}$ between 4 and 10). The protocol here is based on DNA at a concentration of 200 μg/ml.

(5) Make up a stock solution of DNase I (Sigma cat. no. d-1126, electrophoretically purified) at 1 mg/ml in distilled water. Only a very small amount (0.01 ml) of this stock will be needed for each microgram of DNA to be nick-translated.

(6) Add to the vial containing the dried XTPs the following components in the order given: 0.01 ml of the salts mixture; 0.005 ml of DNA (1 μg); distilled water to make up to 0.094 ml.

(7) Pre-incubate at 15 °C for 10 min and then chill in iced water.

(8) Add 0.005 ml (12.5 units) of DNA polymerase I (Grade 1, Boehringer-Mannheim, 550 units/0.1 mg/0.22 ml). The concentration of this enzyme varies from batch to batch. If the specific activity of the enzyme is different from that given above then take about 12 units and change the volume of distilled water so that the total volume of reaction mix at this stage is 0.099 ml.

(9) Dilute the 1 mg/ml stock of DNase I down to 10^{-3} mg/ml in two successive steps of 0.01 ml : 0.99 ml and 0.1 ml : 0.9 ml. Do this just before it is needed.

(10) Add to the reaction mix 0.001 ml of the dilute (10^{-3} mg/ml) DNase I using a very accurate micropipetting system.

(11) Incubate at 15 °C for 1 h. The reaction will proceed linearly for approximately 1½ h.

(12) Add 0.005 ml of the reaction mix to 0.895 ml of a solution containing 50 μl of bovine serum albumin stock (1 mg/ml) and 845 μl water, and add 0.10 ml of 100% TCA. Keep on ice for 15 min. Pass 5 ml of ice-cold 5% TCA through a glass fibre filter (Whatman GF/C, 2.5 cm diameter) and then pass through the reaction mix–serum albumin–TCA sample. Wash the filter by passing through three 5 ml lots of cold 5% TCA. Dry the filter in a 65 °C oven for 20 min and then count in a liquid scintillation system using a toluene-based scintillation fluid.

(13) Take another 0.005 ml sample of the reaction mix, put it directly on another glass fibre filter and count it in a liquid scintillation system, using the same conditions as for the TCA-treated filter. Determine the percentage incorporation of radioactivity into the cold TCA-insoluble fraction. The TCA-treated filter should give between 20 and 60% of the counts obtained from the untreated sample.

This determination of percentage incorporation should be done during the last few minutes of the nick-translation reaction or immediately after the reaction has been slowed by placing the reaction vial on ice.

(14) Stop the reaction by adding to the vial 0.1 ml of water-saturated phenol and mix well with a Pasteur pipette.

(15) Centrifuge hard (about 5000g) for 5 min.

(16) Take the aqueous supernatant and load it directly on to a G50 Sephadex column (1 cm × 35 cm) that has been prewashed with distilled water.

(17) Elute with distilled water, collecting 30-drop fractions. Take 0.005 ml from each fraction and count it in a liquid scintillation system using a tergitol scintillator (PPO 5.0 g, POPOP 50 mg, tergitol 15-s-9 (Union Carbide) 350 ml, toluene 650 ml), or FisoFluor '1' (Fisons Scientific Apparatus). Combine the fractions comprising the first of the two peaks of radioactivity to come off the column, freeze dry them and redissolve the DNA in 0.05 ml distilled water.

The DNA is now ready for use.

The nick-translation protocol given above is one that undoubtedly works and is relatively simple to use. However, as new enzymes, biochemicals, and equipment are developed it will undoubtedly be possible to simplify the protocol still more and reduce the time and cost of the operation. It may therefore prove useful to consult a nucleic acid biochemist or some recent publications on the subject before proceeding with an experiment.

C. Some important comments

The success of a nick-translation largely depends on controlled use of the DNase 1 and the avoidance of other unwanted sources of nuclease activity. There are two main sources of unwanted nuclease activity. The first is the bovine serum albumin used in the salt solution. The second is the DNA polymerase I. BRL now market a nuclease-free bovine serum albumin (cat. no. 5561UA) and Boehringer Mannheim market an endonuclease-free DNA polymerase I (cat. no. 642711).

For those who do not wish to take the trouble to make up their own reagents and who can purchase off-the-shelf enzymes a number of nick-translation 'kits' are now available and are highly recommended, and they are not unduly expensive. An example of such a kit is that supplied by BRL (cat. no. 8160SA).

A most useful account of the nick-translation procedure together with a good list of references can be found in Technical Bulletin no. 80/3 produced by The Radiochemical Centre at Amersham.

VIII. Immunological *in situ* hybridization

Quite recently an interesting and most promising new technique has been introduced into the field of *in situ* hybridization. The technique involves the use of biotin labelled hybridization probes which are then detected in the cytological preparation by histochemical or antibody sandwiching techniques. Of course the great advantage of this approach is that it dispenses with the need for autoradiography, whilst at the same time being just as specific and just as sensitive as more conventional methods that use radioactive probes.

The biotin method has been developed by Langer (1981) and Langer and Ward (1981). It requires the following steps: (1) the synthesis of biotin labelled nucleotides, specifically biotinyl-deoxyuridine 5^1-triphosphate (B-dUTP); (2) the preparation of rabbit anti-biotin antibodies; (3) labelling of a DNA probe with B-dUTP by nick translation; (4) *in situ* hybridization using the labelled probe; (5) treatment of labelled preparations with rabbit anti-biotin; (6) incubation of preparations in goat anti-rabbit immunoglobin g conjugated with fluorescein isothiocyanate (FITC); (7) visualization of the labelled sites on chromosomes by fluorescence microscopy.

Steps (1) and (2) are well beyond the means of most cytology or chromosome laboratories. Steps (3) and (4) follow closely the methods given previously in this chapter. The hybridization is particularly simple, involving incubation at 65 °C with a probe made up in 2× SSC. Steps (5)–(7) are a matter of having the necessary materials and equipment. Details of the method can be found in Langer (1981) and Langer and Ward (1981). It seems virtually certain that this method will become extremely popular, and biotin labelled nucleotides and rabbit anti-biotin are already commercially available (Enzo Biochemicals Inc., see Appendix 1). In principle, there is no reason why *in situ* hybridization should not now be taken right into the classroom or out into the field as a fast and highly effective method for detecting specific DNA sequences with quite simple histochemical procedures.

To be sure, the results obtained by Langer and her associates on polytene chromosomes from *Drosophila* are extremely impressive and show great promise in terms of operational simplicity and cytological resolution.

Finally, it should be said that at the time of writing this book, new developments and advances with the *in situ* hybridization technique are happening almost every day. The scene at the end of 1982 was well stated by Szabo and Ward (1982) in their article entitled 'What's new with hybridization *in situ*'. Researchers interested in using the technique should perhaps begin by reading that article. They should then be able to make use of the advice that we have offered in this chapter in conjunction with all the latest developments in probe construction, hybridization, and detection.

References

Diaz, M. O., Barsacchi-Pilone, G., Mahon, K. A., and Gall, J. G. (1981). Transcripts from both strands of a satellite DNA occur on lampbrush chromosomes of the newt *Notophthalmus*. *Cell*, **24**, 649–659.

Gall, J. G., and Pardue, M. L. (1971). Nucleic acid hybridization in cytological preparations, *Meth. Enzymol.*, **21**, 470–480.

Gerhardt, D. S., Kawasaki, E. S., Bancroft, F. C., and Szabo, P. (1981). Localization of a unique gene by direct hybridization *in situ*. *Proc. Natl Acad. Sci.*, **78**, 3755–3759.

Harper, M. E., and Saunders, G. F. (1981). Localization of single copy DNA sequences on G-banded human chromosomes by *in situ* hybridization. *Chromosoma (Berl.)*, **83**, 431–439.

Harper, M. E., Axel, A., and Saunders, G. F. (1981). Localization of the human insulin gene to the distal end of the short arm of chromosome 11. *Proc. Natl Acad. Sci.*, **78**, 4458–4460.

Hayashi, S., Gillam, I. C., Delaney, A. D., and Tener, G. M. (1978). Acetylation of chromosome squashes of *Drosophila melanogaster* decreases the background in autoradiographs from hybridization with ^{125}I-labelled RNA. *J. Histochem. Cytochem.*, **26**, 677–679.

Lai, E. C., Woo, S. L. C., Dugaiczyk, A., and O'Malley, B. W. (1979). The ovalbumin gene: alleles created by mutations in the intervening sequences of the natural gene. *Cell*, **16**, 201–211.

Langer, P. R. (1981). Localization of biotin labelled hybridization probes by immunological and histochemical methods. PhD thesis, Yale University, New Haven, Conn.

Langer, P. R., and Ward, D. C. (1981). A rapid and sensitive immunological method for *in situ* gene mapping. In *Developmental Biology using Purified Genes*, (D. D. Brown, ed.) Academic Press, New York and London, pp. 647–658.

Lederman, L., Kawasaki, E. S., and Szabo, P. (1981). The rate of nucleic acid annealing to cytological preparations is increased in the presence of dextran sulphate. *Anal. Biochem.*, **117**, 158–163.

Macgregor, H. C. and Mizuno, S. (1976). *In situ* hybridization of 'nick translated'

^{3}H-ribosomal DNA to chromosomes from salamanders. *Chromosoma (Berl.)*, **54**, 15–25.

Malcolm, S., Barton, P., and Ferguson-Smith, M. A. (1981a). The chromosomal distribution of repetitive DNA sequences within the human beta globin gene cluster. *Human Genet.*, **57**, 388–393.

Malcolm, S., Barton, P., Murphy, C., and Ferguson-Smith, M. A. (1981b). Chromosomal localization of a single copy gene by *in situ* hybridization – human beta globin genes on the short arm of chromosome 11. *Ann. Human Genet.*, **45**, 135–141.

Maniatis, T., Jeffrey, A., and Kleid, D. G. (1975). Nucleotide sequences of the rightward operator of the phage lambda. *Proc. Natl Acad. Sci.*, **72**, 1184–1188.

Pukkila, P. J. (1975). Identification of the lampbrush chromosome loops which transcribe 5 s ribosomal RNA in *Notophthalmus (Triturus) viridescens*. *Chromosoma (Berl.)*, **53**, 71–89.

Rigby, P. W. J., Dieckmann, M., Rhodes, C., and Berg, P. (1977). Labelling of deoxyribosenucleic acid to high specific activity *in vitro* by nick translation with DNA polymerase I. *J. Mol. Biol.*, **113**, 237–251.

Singh, L., Purdom, I. F., and Jones, K. W. (1977). Effect of different denaturing agents on the detectability of specific DNA sequences of various base compositions by *in situ* hybridization. *Chromosoma (Berl.)*, **60**, 377–389.

Szabo, P., Elder, R., and Uhlenbeck, O. (1975). The kinetics of *in situ* hybridization. *Nucleic. Acids Res.*, **2**, 647–653.

Szabo, P., Elder, R., Steffensen, D. M., and Uhlenbeck, O. C. (1977). Quantitative *in situ* hybridization of ribosomal RNA species to polytene chromosomes of *Drosophila melanogaster*. *J. Mol. Biol.*, **115**, 539–563.

Szabo, P., and Ward, D. C. (1982). What's new with hybridization *in situ*? *Trends in Biochem. Sci.*, **7**, 425–427.

Wahl, G. M., Stern, M., and Stark, G. R. (1980). Efficient transfer of large DNA fragments from agarose gels to diazobenzyloxymethyl paper and rapid hybridization using dextran sulphate. *Proc. Natl Acad. Sci.*, **76**, 3683–3687.

Wimber, D. E., and Steffensen, D. M. (1973). Localization of gene function. *Annu. Rev. Genet.*, **7**, 205–223.

9

Autoradiography

The object of this chapter is to introduce a method of autoradiography that works. It should be borne in mind, however, that there are many variations of this basic method, and each experienced investigator is likely to have his or her own particular brand of witchcraft. Each step in the following procedure is reasonable in the sense that with proper care and attention it leads to the production of clean, clear, and generally good autoradiographs of animal or plant cells that have been labelled with tritiated compounds. The whole pattern of the process as outlined here represents a combination of suggestions and methods derived from many people who have first hand experience of the practical application of the technique.

The isotope most frequently employed in autoradiography is tritium (^{3}H). It is an unstable isotope of hydrogen emitting weak β radiation. The half-life of tritium is 12.26 years. The mean particle energy is 0.0186 MeV, which compares with 0.156 MeV for carbon-14 and 1.710 MeV for phosphorus-32. The approximate range in dry biological materials and nuclear track emulsions of β particles from tritium atoms is about 1 μm, although 1–2% of the β particles do have ranges of up to 3 μm.

Although tritium is not classed as a dangerous isotope it should nonetheless be handled with care, and rules relating to proper laboratory practice and safety should be closely followed.

The tritium-labelled compounds most commonly used with cell nuclei and chromosomes are amino acids, nucleotides, nucleoside triphosphates, and nucleic acids. Many organic and inorganic compounds labelled with tritium can be obtained from Amersham Radiochemical Centre in the UK and from New England Nuclear in the USA. Investigators should consult the catalogues of these companies for information on availability and prices. Tritium labelling of nucleic acids can be accomplished either *in vivo* or *in vitro*. The *in vivo* approach generally involves injecting the labelled precursor into the animal followed at an appropriate time by the removal of tissue and the production of chromosome preparations. The *in vitro* approach can be at two distinct levels. On the one hand, pieces of animal tissue can be cultured *in vitro* in a medium containing a labelled compound; on the other hand, a nucleic acid can be isolated, purified, and then labelled *in vitro* in a biochemical reaction involving a nucleic acid polymerase. The culture technique has been extensively used in the study of lampbrush and polytene chromosomes.

Nucleic acids can be labelled biochemically in several ways, but the simplest and therefore the most commonly used methods are through the production of a complementary RNA (cRNA) employing RNA polymerase from *E. coli*, the 'nick-translation' of DNA employing deoxyribonuclease I and DNA polymerase I, and the controlled excision and $5'-3'$ polymerization of DNA with T4 polymerase. Simple protocols for the first two of these methods and appropriate references for the third are given in Chapter 8.

Amino acids and nucleotides can be labelled with tritium in one or more positions in the molecule where hydrogen atoms normally occur. When designing an experiment involving tritiated uridine or thymidine, for example, careful account should be taken of the positions of the hydrogen atoms that have been replaced by tritium and of what will happen at these positions as the nucleotides pass along any metabolic pathways before being incorporated into RNA or DNA.

Other factors that need thinking about when designing an experiment that involves autoradiography are the concentration of the isotope and the specific activity (usually expressed as disintegrations per minute per microgram) of the labelled compound that is to be used. Tritiated uridine or thymidine is usually supplied at a concentration of 1 mCi/ml of aqueous solution. The specific activity may be anything from 1 to 30 Ci/mM. This latter factor will obviously affect experimental design in so far as with the 'hotter' compound a shorter autoradiographic exposure is likely to be needed than with the less 'hot' compound because more tritium will be incorporated into tissue per mole of precursor and the tissue will therefore be more radioactive. Specific activity is not usually a problem with nick-translated nucleic acids, cRNAs, or T4 polymerase products. With these it is easy to attain levels of between 10^7 and 10^9 disint./min, and this is very hot by any standards.

Yet another factor that should be taken into account when designing experiments with radioisotopically labelled compounds is the cost of the labelled compound. Tritium-labelled compounds are relatively inexpensive as compared, for example with ^{14}C-labelled compounds. Nick translations are appreciably cheaper than cRNA preparations.

I. Autoradiographic emulsions

Photographic or nuclear track emulsions consist of silver halide crystals suspended in a gelatin matrix. When these crystals are exposed to radiation either from a light source or from a radioactive source, or when they are subjected to mechanical friction, they become sensitized in such a way that they are reducible to metallic silver upon treatment with a photographic developer. This sensitization is often referred to as latent image formation.

The liquid nuclear-type emulsions manufactured for autoradiography are specifically designed so that by microscopic examination of the positions of silver grains it is possible to identify sources of ionizing radiation in a specimen underlying the emulsion. The emulsions are designed with characteristics of fine grain, uniform grain size distribution, a very high ratio of silver grains to gelatin, and a very low background 'fog' level. Emulsions are available with several different sensitivities. Some of the most commonly used and their characteristics are listed in Table 9.1. Later in this chapter some details and a protocol will be given relating to the use of 'stripping plates' in autoradiography.

Table 9.1 Commonly used nuclear track emulsions for autoradiography

Manufacturer	Type	Mean crystal diameter (μm)	Properties
Ilford	K2	0.20	A moderately sensitized emulsion suitable for most tritium autoradiography
Ilford	K5	0.20	More highly sensitized to all charged particles of any energy
Ilford	L4	0.13	A fine grain emulsion sensitized to all charged particles of any energy; particularly useful for electron microscope autoradiography
Kodak	NTB	0.34	A highly sensitive emulsion recording all charged particles; particularly suitable for γ radiation emitters
Kodak	NTB2	0.26	A good general purpose emulsion for tritium autoradiography, recording particles with energies of less than 0.2 MeV
Kodak	NTB	0.29	Suitable for recording particles with energies of less than 30 keV; useful for α particle emitters

Further information on Ilford and Kodak products can be obtained from the manufacturers; their addresses and telephone numbers are listed in Appendix 1

One of the most commonly used nuclear track emulsions in chromosome work is Kodak NTB2 and the instructions that follow relate specifically to this material. However, they are equally applicable to other emulsions listed in Table 9.1.

Nuclear track emulsions are expensive and delivery time on most of them outside the country of manufacture is usually quite long. It is therefore important to know how to store them, how to handle them, and how long they will last. They must always be stored under refrigeration, but *never* frozen. Emulsions that have been kept at room temperature for long periods or frozen will probably be totally useless. Nuclear track emulsions have limited useful lives in so far as unacceptably high levels of background fog occur after long periods of storage. However, it has to be said that the shelf lives given by manufacturers are exceedingly pessimistic. With careful handling and storage NTB2 remains in an entirely satisfactory condition for 5 years or more, and even when the background fog reaches critical levels it can often be reduced by employing the peroxide treatment given later in this chapter. Moreover, the NTB emulsions can certainly be used several times over with only slight increase in background after each round of melting and dipping. Of course it should be added that the very best results are most likely to be obtained using new emulsion that has not previously been handled.

NTB2 is best used diluted 1 : 1 with water. With this consideration in mind the following procedure is recommended for dealing with a new 4 fl. oz. (118.29 ml) bottle of NTB2 immediately upon receipt from the suppliers. It involves checking the emulsion, diluting it, and then distributing it into aliquots that are suitable for subsequent storage and use.

(1) Set up the darkroom as specified later in this chapter (Section II), with safelight, hot plate, water bath at 45°C, slide drying rack, and fan. The water bath should have just enough water in it to allow a 500 ml beaker containing 200 ml of water to stand upright in it.

(2) Prepare to work in total darkness, which means having everything to hand and in predetermined positions on the bench.

(3) Take the following items into the darkroom and lay them out carefully and systematically on the bench: (a) the sealed and unopened package of emulsion; (b) three very clean microscope slides in a glass staining rack; (c) one 250 ml Erlenmeyer flask, scrupulosuly clean and containing 200 ml of carefully filtered glass-distilled water and capped with foil; (d) one empty and very clean 500 ml beaker; (e) 22 small glass vials measuring about 4 cm × 2 cm with plastic snap-on tops, and again very clean; (f) two good quality cardboard boxes, one of which will fit comfortably into the other, and the smaller of which will accommodate all the 22 small glass vials.

(4) Turn out all the lights and check that the room is quite dark.

(5) Open the box of emulsion, remove the plastic bottle, and stand it in the water bath. Stand the flask of distilled water in the water bath and put the 500 ml beaker on the hot plate to warm.

(6) Place the rack of slides on the hot plate to warm.

(7) Wait for 1 h. If the darkroom has double doors or a good light trap around the door, then leave and get on with something else while waiting for the emulsion in the bottle to melt.

(8) After 1 h remove the bottle of emulsion from the water bath and invert it several times to check that the contents are fluid and well mixed. Do not shake vigorously. Remove the top slowly and gently and stand the bottle again in the water bath.

(9) Dip each of the three slides, one at a time, into the bottle of emulsion, taking special care not to rub the slides against the mouth of the bottle and not to let fingers come in contact with the emulsion.

(10) Stand the slides in the drying rack.

(11) Put the top back on the bottle of emulsion, leaving the bottle standing in the water bath.

(12) Start the fan so that it blows gently over the newly dipped slides from a distance of about 50 cm.

(13) Wait for 1 h.

(14) Remove the slides from the drying rack and develop, fix, wash, and dry them exactly as described later in this chapter (Section VIII), taking care not to get photochemicals anywhere near the emulsion or other equipment on the darkroom bench.

(15) Take the slides away for immediate examination, leaving the darkroom securely locked and the bottle of emulsion capped and still standing in the water bath.

(16) Place a slide on a microscope that has a 100× oil-immersion objective. Put a drop of immersion oil directly on to the slide, focus on the level of the emulsion and bring the background silver grains clearly into view. As a general rule, the emulsion may be considered to be satisfactory if there are fewer than 50 silver grains visible per field of view when using a 100× objective. If the grain density is appreciably higher than this and certainly if it is higher than 100 per field then return directly to the darkroom, cap, box, and seal the package of emulsion and return it to the suppliers with a request for a replacement and a letter describing exactly how the test was carried out. Both Kodak and Ilford will normally replace faulty emulsions immediately and without charge if they are satisfied that the customer handled the product correctly and carried out the test properly.

If the grain density is satisfactory then return to the darkroom and proceed as follows.

(17) Switch on the red safelight.

(18) Pour the whole contents of the bottle of emulsion into the prewarmed 500 ml beaker and stand the beaker in the water bath.

(19) Fill the plastic emulsion bottle with prewarmed distilled water from the Erlenmeyer flask.

(20) Put the cap on the plastic bottle and shake the bottle to mix the remaining emulsion with the distilled water.

(21) Remove the cap from the bottle and pour the whole contents into the 500 ml beaker.

(22) Pick up the beaker and thoroughly mix the water with the emulsion by swirling them around the bottom of the beaker for several minutes. Do not use a stirring rod or magnet stirring bar as these may cause mechanical exposure of the emulsion. Mix only by swirling.

(23) When satisfied that the diluted emulsion in the beaker is well mixed, dispense it into all of the 22 glass vials, filling each vial almost to the top.

(24) Finally, cap the vials securely, place them in the smaller of the two boxes, put the top on the box and place it inside the larger box. Seal the outer box with tape and mark it to indicate its contents, the date of receipt of the emulsion, batch number, and manufacturer's date of expiry.

The emulsion is now ready for use. One vial may be expected to coat up to 50 slides. The box of diluted emulsion should be stored at 3–5°C in a refrigerator.

II. Darkroom

The darkroom used for autoradiography should be completely light tight and should be securely lockable from inside and outside. It should be equipped with a Kodak Wratten Series 1 red safelight with a low wattage bulb, a constant temperature water bath containing clean distilled water and set to hold 45°C, a slide rack capable of holding 20 or more slides in an upright standing position with only the bottom edges of the slides in contact with the rack, a table-top fan powered by a brushless motor, a hot plate with a surface temperature of about 60°C, and a substantial area of clear bench. All equipment should be easily usable in total darkness. Nothing should be kept in the darkroom that is capable of being knocked from the bench or spilled. Pay attention to the lighting system of the darkroom. If this is of the strip fluorescent type then ascertain that the tubes do not glow spontaneously after switching off. If they do, remove them. Also, remove or mask all warning or pilot lights on water baths and hot plates. And beware of water baths with transparent plastic liners. The pilot light may be effectively masked from the outside but will glow brightly inside the bath!

III.   The dipping chamber

The chamber that will contain the emulsion and into which the slides will be dipped must be constructed so as to hold the minimum quantity of emulsion that is needed for effective dipping, it must be long and wide enough to accommodate all of one standard microscope slide, and it must be easily cleanable. The best dipping chambers are made of glass and measure approximately 8 cm × 3 cm × 0.7 cm. However, such chambers are not available commercially and many autoradiographers have found it simpler and equally effective to construct dipping chambers from Perspex (Plexiglass) or to use the small plastic containers that are available for mailing microscope slides. ('Cytomailers', American Scientific Products, cat. no. S7775). An arrangement should be made such that the dipping chamber can be stood firmly in the water bath in an upright position and immersed to within 0.5 cm of its top. A small plastic funnel should be available to assist in pouring the prediluted and melted emulsion from a glass storage vial into the dipping chamber. The position of the dipping chamber in the water bath should be such that it can easily be located in total darkness.

IV.   Slide storage boxes for autoradiographs

Several types of storage box are available from scientific suppliers in all parts of the world. The most suitable ones are made of black or grey plastic (American Scientific Products, cat. no. M-6295-box, or R. A. Lamb, cat. no. E-38). They accommodate about 25 slides and have neatly fitting lids. They should be dust free and dry. Autoradiographs must be stored dry during exposure and for this reason a small amount of dessicant is included in each box. A glass vial of about the size of a liquid scintillation vial (5 cm × 2.5 cm) is half filled with dessicant (drierite, silica gel, or equivalent). The neck of the vial is loosely plugged with cotton wool. The vial can then be capped securely until required for use. A microscope slide is placed in about the eighth position of the storage box. This creates a compartment in which the vial of dessicant can be placed. The remainder of the box is available for prepared autoradiographs that are to be left to expose.

V.   Subbed slides

Preparations that are to be autoradiographed should always be made on subbed slides so as to prevent any possibility of slipping of the nuclear track emulsion during development, fixing, washing, or staining. Slides to be subbed should first be thoroughly cleaned in strong acid, and then rinsed copiously in hot water and distilled water. They should then be dipped directly in subbing solution, allowed to drain, and oven dried at 65°C for

12h. Slides can be subbed in large batches and stored indefinitely in the boxes in which they are received from the supplier.

The subbing mixture consists of powdered gelatin, 0.1 g; chrome alum, 0.01 g; and water 100 ml.

VI. Checklist of equipment and protocol for preparing autoradiographs

(a) Safelight
(b) Water bath
(c) Hot plate
(d) Slide drying rack
(e) Fan
(f) Slide storage boxes
(g) Dessicant vials
(h) Dipping chamber with provision for holding it in the water bath
(i) Small plastic funnel
(j) Slides to be dipped in glass rack
(k) Two new clean slides in a separate rack
(l) Two stout cardboard boxes
(m) Adhesive tape
(n) Marker pen, green or black (not red)
(o) Paper towels or laboratory tissues
(p) Box of vials of emulsion
(q) A roll of aluminium foil

When all these items have been assembled, lock the darkroom and proceed as follows:

(1) Place the rack bearing the slides that are to be dipped and the rack with two clean slides on the hot plate and cover them with foil. This prewarms the slides and is recommended because repeated dipping of cold slides into emulsion at 45°C will eventually lower the temperature of the emulsion and change its consistency.

(2) Take a vial of emulsion from the stock box, check that it is securely capped, and immerse it in the water bath. Leave it there for at least 30 min. During this time it is perfectly reasonable to turn out the safelight and leave the darkroom, locking it securely to prevent mistaken entry by other persons.

(3) When the emulsion has melted, remove the vial from the water bath, dry the outside with a clean tissue, uncap it, and with the aid of the small funnel pour the emulsion into the dipping chamber. Fill the chamber to the top. The capacity of the storage vial should about equal that of the dipping chamber.

(4) Take one of the clean, prewarmed slides and lower it as far as possible into the dipping chamber. Keep a firm hold of it and do not let fingers come

in contact with the emulsion. Hold it there for 10 s and then draw it out quite swiftly and smoothly. Allow the slide to drain for a few seconds against the edge of the dipping chamber and then stand it in the drying rack.

(5) Do the same with the second clean slide.

(6) Do the same with each of the experimental slides. Take note of the order in which the slides are dipped and the order in which they are stood in the drying rack. Work slowly and uniformly.

(7) When all the slides have been dipped, place the drying rack about 50 cm in front of the fan and switch the fan on.

(8) Pour the emulsion from the dipping chamber back into the storage vial. Cap the vial and put a piece of tape around it to signify that it has been once used. Put the vial back in the storage box and close the box securely. Generally tidy up in preparation for subsequent steps. The emulsion on the dipped slides will take about 1 h to dry.

(9) After an hour or more place the dipped slides into the plastic storage boxes. Put the tops on the boxes, tape them securely, mark them to indicate their contents, place them inside the cardboard box, and tape and label that box indicating the date on which the dipping was carried out.

(10) Return the stock box of emulsion to the fridge, and put the box of coated slides where they will not be interfered with.

Warning: Adhesive tape of the kinds that are useful for taping boxes of autoradiographs give off considerable amounts of light from static discharge when a piece of tape is being stripped from a roll or when the tape is being removed from around a box. This light can cause unwanted exposure of nuclear track emulsion.

VII. Exposure

The correct exposure time for autoradiographs has to be determined empirically. It is therefore usual to divide preparations into two groups: those that are not so good cytologically and can be used as 'test' preparations to determine the best exposure time, and those that are good and will not be developed until the test preparations have indicated that exposure has been adequate. The number of test preparations will depend on the total number of preparations that are available and the general nature of the experiment, but as a rule it is best to have at least 5 test preparations so that good assessment of optimal exposure time can be made before committing the best and definitive slides of the experiment to the irrevocable step of development. There are no hard and fast rules about optimum exposure. The objective is simply to see enough silver grains in the preparation to show where incorporation or binding of the labelled compound has taken place, but not so many silver grains that the cytological detail in the underlying preparation is obscured.

VIII. Development

Where there are fewer than 10 slides, development can best be carried out in Coplin jars. With more than 10 slides use staining racks and dishes. Only the Coplin jar method is given here.

All solutions and the water in which the slides will be washed after development and fixing must be at 20°C. Nuclear track emulsions are very sensitive to changes in temperature when they are wet and they will quickly wrinkle, move, or loosen from the slide if they are sequentially subjected to solutions at different temperatures.

Collect together in the darkroom the following items:

(a) The box of slides to be developed
(b) One empty and clean Coplin jar
(c) One Coplin jar full of Kodak D19 developer made up freshly to half-normal strength
(d) One Coplin jar of distilled water
(e) One Coplin jar full of fixer
(f) A timer with an alarm bell or buzzer

Set the Coplin jars out on the bench with the empty one on the left, then the developer, then the water, and then the fixer on the right. Check again that they are all at 20°C. Set the timer to 2¼ min. Turn out the lights and lock the darkroom door. Open the box of slides and remove those that are to be developed. Place them carefully in the empty Coplin jar. Close and retape the box of slides and replace it in the cardboard storage box. Pick up the Coplin jar containing developer and pour its contents into the jar with the slides. Start the timer. When the bell rings after 2¼ min (total development time is 2½ min allowing for pouring time) pour off the developer and pour in the distilled water. Immediately pour off the distilled water and pour in the fixer. Wait a further 2 min and then turn on the lights. Leave in fixer for a total of 5 min. Pour off the fixer and wash the slides in at least five changes of distilled water, each of about 2 min duration. Place the slides in 0.5% formaldehyde solution for 10 min. Wash again in at least five 2 min changes of distilled water. Stand the slides in air to dry. Remember it is most important to have all the distilled water and the formaldehyde solution used in washing and hardening steps at 20°C. After drying the preparations can then be stained with any of a number of dyes, some of which are mentioned below.

Always use freshly prepared photographic chemicals for developing autoradiographs so as to get the best and most reproducible results. D19 should be made up to half strength from the crystalline chemicals immediately before use. It is convenient to weigh out these chemicals and keep them available in stock lots that will make about 1 Coplin jar-full of solution (e.g. 1.1 g component A and 17 g of component B to be dissolved in distilled water to make a total volume of 225 ml when required for use). Never use a stop bath between development and fixing as this sometimes results in the formation of small bubbles beneath the emulsion. Do not use rapid (ammonium) fixers or fixers containing acid hardeners since they can result in loss of silver grains. Kodak Photostat fixer no. 4 is entirely suitable.

Other developers that can be used for NTB emulsions are Kodak Dektol (1 : 1 with distilled water) for 1½ minutes or Kodak D170 (full strength) for 4 min.

IX. Autoradiography with stripping plates

Kodak AR10 stripping plates are most useful when dealing with material that has been embedded in wax or plastic and sectioned at more than 2 μm. Dipping emulsions may not be suitable for such specimens because they build up in pools around the steep contours of the section or the material within it, and accordingly the thickness of the emulsion varies widely from place to place across the specimen. What follows is a protocol that is appropriate for wax sections 5–10 μm thick mounted on subbed slides, dewaxed with xylene, passed through a graded series of alcohols to distilled water, and taken into the darkroom in a Coplin jar of distilled water.

The following equipment and materials should first be assembled:

(a) Safelight, as for NTB
(b) Slide drying rack
(c) Fan
(d) Slide storage boxes with dessicant, etc.
(e) Adhesive tape
(f) A new single edge razor blade
(g) Pair of clean fine-pointed forceps
(h) Package of AR10 plates: these are normally stored in the refrigerator and the package should be removed and allowed to warm to room temperature for about 1 h before use
(i) A large glass finger bowl filled with very clean distilled water to just below the brim: the water surface should be free from dust and grease and the edges of the bowls must be dry

The darkroom and the water in the bowl should be at 25°C. AR10 is almost impossible to handle under very dry conditions due to its tendency to pick up static charge. The humidity of the darkroom should therefore be kept high.

When all these requirements have been met, proceed as follows:

(1) Lock the darkroom, turn out the lights, and turn on the safelight.

(2) Remove an AR10 plate from its box and unwrap it.

(3) Determine which side of the plate carries the emulsion by examining the plate close to the safelight. One side of the plate is obviously uncoated glass and the other is equally obviously coated with emulsion. The *sensitive surface* of the emulsion is that which is *outermost*. It must on no account be touched with fingers.

(4) With a razor blade score the emulsion across the width of the plate about 2 mm in from the edge and then score it lengthwise along the midline of the plate. Then score it across the width of the plate at intervals about equivalent to a little more than the width of a microscope slide. This cuts the emulsion up into pieces that are a little larger than slide-sized.

(5) Hold the plate near to the finger bowl and with the forceps grasp the edge of a strip of emulsion close to the midline score. Peel the strip back gently (rapid peeling may produce static discharge between slide and emulsion and consequent risk of exposure), draping it over the edge of the bowl, and as it finally comes free from the plate quite rapidly throw it on to the surface of the water in the bowl with its sensitive surface downwards. If the emulsion tends to curl up as soon as it is freed from the plate, then conditions are too dry. An increase in the humidity of the room should solve the problem. The strip of emulsion should now be lying face downwards and should be quite free from wrinkles, folds, or tears.

(6) Allow about 2 min for the emulsion to swell on the water surface, then take a slide preparation side uppermost, pass it beneath the emulsion, and lift it up so that the emulsion drapes over its surface. Twist the slide to and fro in the air so that the edges of the emulsion come to wrap around the back of the slide. Stand the filmed slide in the drying rack in front of the fan and leave it to dry for about 1 h.

(7) When the slides are completely dry put them in the storage boxes, tape and mark everything securely, and place in a refrigerator at 3–5°C to expose.

As with NTB emulsions, exposure time must be determined empirically. Development, fixing, and staining of AR10 is the same as for NTB2.

X. Removal of autoradiographic background with peroxide

The nuclear track emulsions used in autoradiography always include some sensitized silver halide crystals and these appear as undesirable 'background' grains in the final autoradiograph. Treatment with hydrogen peroxide before exposure of the emulsion to radiation from the preparation oxidizes the sensitized silver halides, removes the latent image, and therefore reduces the number of background grains in the final autoradiograph.

The procedure for peroxide treatment of slides is as follows. A piece of standard laboratory Kleenex tissue 6 cm × 6 cm is folded into a small package measuring about 2 cm × 1 cm and tied with a small piece of Sellotape (Scotch tape). The tissue is then soaked in a 6% solution of hydrogen peroxide. Excess solution is shaken from the tissue which is then placed in the compartment of the black plastic slide box normally occupied by the vial of dessicant. The dried, newly filmed autoradiographs are then placed in the other compartment of the box, and the box is closed and left to stand for 1–2 h. In total darkness, the box is then opened, the tissue removed, the compartment blotted dry with a piece of lint-free tissue, and the uncapped vial of dessicant is placed in the compartment. The box is then closed, sealed with tape, placed inside a light-tight cardboard box, and left to expose.

XI. Staining of autoradiographs

There are several methods for staining autoradiographs. All of them involve staining after developing, fixing, and drying the final autoradiograph. Five adequate recipies are given below. In all these cases the dried autoradiograph is directly immersed in the staining solution, stained, washed, and air dried. Stained preparations can be examined directly with oil-immersion objectives and no coverglass, or they can be permanently mounted with coverglass using a suitable synthetic mountant. Canada balsam is not suitable since it has a tendency to become acid with age, in which case it completely destroys the silver grains in the underlying preparation.

(a) 1% methylene blue in 1% borax (sodium tetraborate) solution for 5 min. Rinse quickly in distilled water.

(b) 0.15% methyl green at pH 6 in phosphate buffer (0.01 M) for 10 min. Rinse in buffer briefly and then in distilled water.

(c) 0.04% toluidine blue in 0.001 M phosphate buffer at pH 6 for 10 min. Rinse in distilled water.

(d) Giemsa stain. Place 0.5 g of commercial Giemsa powder in 33 ml of glycerol. Hold at 60°C for 2 h. Add 33 ml of methanol and stir well. This stock solution can be kept for months. Otherwise, use ready made-up Giemsa stain (e.g. Gurr's R66). For staining, place the slides in a Coplin jar containing about 60 ml of 0.01 M phosphate buffer, pH 7. Add 5–10 ml of Giemsa stain stock solution and mix it into the buffer by pumping in and out with a Pasteur pipette. Leave to stain for 15–60 min or more depending on the staining intensity required. After addition of the Giemsa to the buffer a metallic film will form on the surface of the staining solution. The slides must on no account be drawn through this film. Therefore it is essential that the stain solution should not be poured off from the Coplin jar but that it be flushed out with a stream of distilled water.

(5) Coomassie Blue. Pass the slides quickly through 50% methanol and into 0.1% Coomassie Blue R250 in 50% methanol and 10% acetic acid for 10 min. Rinse briefly in 50% methanol; wash in distilled water.

Additional sources of information

Tritium Labelled Molecules in Biology and Medicine, by Ludwig E. Feinendegen. Chapter 4, Autoradiography. Academic Press, New York and London (1967).

Medical Cytogenetics and Cell Culture, 2nd ed, by Jean H. Priest. Chapter 9, Light Microscope Tritium Autoradiography. Lea and Febiger, Philadelphia (1977).

Techniques of Autoradiography, by Andrew W. Rogers. Elsevier Publishing Company, Amsterdam, London and New York (1969).

Kodak Materials for Autoradiography, Kodak Technical Information Sheet no. 1977/EJH. This contains a number of useful references relating to the handling and application of different emulsions. See Appendix 1 for Kodak address.

Ilford Product Information Sheet entitled 'Nuclear research materials – a range of materials for autoradiography and nuclear physics research' (1982). See Appendix 1 for Ilford address.

10

Measuring nuclear or chromosomal DNA

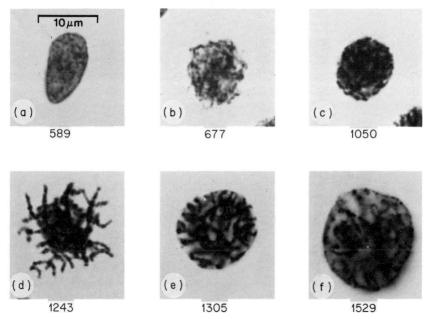

Photomicrographs of nuclei in a Feulgen-stained preparation of a single ovary from the Pacific tailed frog, *Ascaphus truei*. All nuclei were on the same slide and all were photographed under the same conditions and the photographs processed alike. The number beneath each nucleus is the Feulgen dye content of that nucleus in arbitrary units as measured photometrically. The nucleus in (a) is a somatic fibroblast nucleus with a DNA content that is likely to be close to the 2C amount for the species. All other nuclei are in meiotic prophase, ranging from leptotene in (b) through to early diplotene in figure (f). The increase in Feulgen dye content reflects the increase in amount of nucleolar DNA resulting from amplification of the genes for ribosomal RNA [Macgregor, H. C., and Kezer, J., *Chromosoma (Berl.)*, **29**, 189–206 (1970).]

Whatever the objectives of looking at chromosomes there nearly always comes a point at which the cytologist wants to discuss results and observations in molecular terms, and this usually means thinking in terms of actual amounts of nuclear or chromosomal DNA. There are three main ways of measuring nuclear DNA. The first involves biochemistry. A nuclear preparation is made and the number of nuclei per millilitre are counted either by microscopy with a haemocytometer or using a Coulter counter. DNA is then extracted from a known number of nuclei and purified to the point at which its absorption spectrum and absorbance is as free as possible from interference by such things as proteins, polysaccharides, and RNA. At each step in the purification procedure care is taken to monitor loss of DNA so that this may be compensated out in the final analysis. The amount of purified DNA extracted from a known number of nuclei is determined by spectrophotometry using light of 260 nm wavelength. An example of the application of this technique for the determination of C values of certain plethodontid salamanders can be found in Mizuno and Macgregor (1974). Other more complex biochemical methods involve examining the reassociation kinetics of denatured chromosomal DNA and estimating the size of the total genome from its complexity and structure in relation to the more simple genome of a bacterium such as *Escherichia coli*. Such methods are beyond the scope of this book and they will not be considered further.

The other two principal methods for DNA measurement are both cytophotometric and provide values that have to be related to measurements made on nuclei or chromosomes with known DNA contents in order to be able to express results in terms of actual amounts of DNA. One of the most popular and undoubtedly the fastest method of measuring nuclear and chromosomal DNA is by flow cytometry. The method allows rapid analysis of large numbers of nuclei or of chromosomes if they can be isolated and prepared in sufficient quantities from cultures of actively dividing cells. The nuclei are usually stained with a fluorescent dye such as mithramycin and detected and measured with standard electronic instrumentation and a photomultiplier tube (PM) as they pass through a focused laser beam (Melamed *et al.*, 1979). Flow cytometry has been widely applied for DNA measurements over the past few years. It is undoubtedly a useful and accurate way of measuring nuclear DNA, but it is of little use when it comes to measuring components within nuclei or relating measurements to nuclear form.

I. Microdensitometry

The oldest and in some senses the most dependable method of DNA measurement is by Feulgen microdensitometry, and in what follows we shall concentrate wholly on this particular approach. The tissue is fixed, and stained with the Feulgen reaction before or after squashing, splashing, sectioning, etc. The Feulgen reaction (Feulgen and Rossenbeck, 1924) entails mild hydrolysis of nuclear DNA followed by staining of the unmasked aldehydes with Schiff's reagent (Kasten, 1960, 1970). It is now widely accepted that under carefully controlled conditions of hydrolysis and staining, Feulgen is both specific for DNA and stoichiometric (Swift, 1955; Kasten, 1960; Deitch, 1966). Once the specimen has been stained the amount of bound dye is measured with a microdensitometer fitted to a normal research microscope, employing light of a wavelength near to the absorbance maximum of the dye being used. In recent years there have been quite spectacular advances in the design of instrumentation for microdensitometry. Nowadays, almost any cytological object can be measured, and the kinds of questions that can be answered are likely to be limited only by the sophistication of the equipment that is at hand.

Persons starting out on a programme of microdensitometry with the Feulgen reaction should first read very carefully through some of the older literature in the field, literature that explains the fundamentals of the approach. Microdensitometry needs good microscopy, good cytochemistry, and good evaluation. It is therefore essential to understand both what has to be done and why it has to be done. Chapters by Swift (1955) and Pollister *et al.* (1969) are particularly useful. It matters little that these papers were written long before the more sophisticated and sensitive measuring devices of the present day were available. Even the very best microdensitometers of the 1980s can generate the most spurious and misleading results if they are not sensibly used with a strong appreciation of the basic elements of the technique. It is for this reason that we have not attempted here to provide a full working guide to the technique. Such a venture would need an entire book in itself. Instead we are outlining the options, providing a sound working protocol for the Feulgen reaction, and offering some advice on sources of information and systems of instrumentation.

The simplest procedure in microdensitometry applies to a specimen in which the distribution of the absorbing material is perfectly uniform. Imagine a cell nucleus in the form of a perfectly round, flattened cylinder with all of its upper surface parallel to all of its lower surface and the whole thing evenly stained with Feulgen dye, the intensity of staining being strictly a function of the amount of DNA in the nucleus. The microdensitometer is first arranged so that the area over which it measures is filled by an area of the preparation that includes no light-absorbing material. The only materials, other than lenses, that interrupt the light path between the light source and the photomultiplier tube are the microscope slide, the coverglass, and the medium in which the specimen is mounted. The wavelength of the light is set to the absorption maximum of Feulgen dye (546 nm) and the meter associated with the PM tube is set to a full scale deflection representing 100% transmission. The object is then moved into position to occupy the whole of the field of measurement. The percentage transmission is then noted. The area and thickness of the specimen are determined. The area of the field over which the absorption was measured is known. It is then possible to express the total Feulgen dye content of the whole specimen in arbitrary units of absorption and to compare this with the total Feulgen dye content of any other uniformly stained specimen of similar geometric shape stained and measured under identical conditions. The principle is exactly that of spectrophotometry, with the cell nucleus acting like a parallel-sided cuvette, the absorbing molecules randomly arranged as in a perfect liquid, and the laws of absorption holding true. Unfortunately such a state of affairs never exists. Nuclei are never perfectly round, they are never perfectly flat, and they rarely stain homogeneously. So what do we do?

In effect, the problem of obtaining meaningful measurements of the dye contents of odd-shaped and patchily stained cytological specimens reduces to one single factor – inhomogeneity of the field of measurement. In the same sense as a cell nucleus will present an inhomogeneous image on account of the uneven distribution of chromatin, so a set of chromosomes occupying the measuring area of a densitometer will present an inhomogeneous image on account of lack of staining of the material between individual chromosomes. So if a way can be found to overcome the errors introduced by this heterogeneity of distribution of dye within the measuring area, then measurement of the dye content of any object will only be limited by the rules of good light microscopy and the sensitivity of the PM equipment.

Two methods exist for coping with the problem. The first of these was devised by Patau in 1952. It is highly ingenious and if properly applied it allows very accurate measurements to be made even on very small specimens. The principle is as follows. Figure 10.1 shows two absorption curves, one from a nucleus that is homogeneously stained (curve A) and the other from a nucleus of the same size and dye content where the dye is arranged in a few strongly staining clumps interspersed with clear areas (curve B). The peak of curve B is noticeably depressed. In figure 10.2 the same data are reproduced, but here the absorptions of the homogeneous nucleus are plotted against those from the inhomogeneous one. At higher absorptions the disparity between the two curves increases.

Where any two areas containing the same pigment or dye are to be compared, it is easy to see that each may be measured at two or more wavelengths and the ratios of the absorbances determined. Where the ratios are different a distributional error is indicated. Furthermore, the extent of difference is easily computed and a correction for it can be applied. This effectively means that any irregular areas such as a part of a Feulgen-stained chromosome or a patchily stained nucleus can easily be measured with no more than an intelligent regard to heterogeneity in the distribution of the

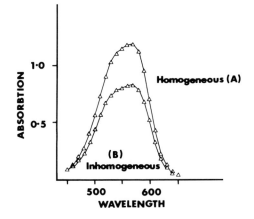

Figure 10.1 Absorption curves for Feulgen-stained nuclei. Curve A, homogeneously stained nuclei; curve B, inhomogeneously stained nuclei having the same DNA content and general size as those represented in curve A. The depression of the absorption curve for inhomogeneous nuclei is used as a basis for estimating the error in absorption measurement introduced by irregular distribution of Feulgen-stained material in the nucleus as well as the general irregularity produced by including the whole of a cell nucleus with some empty space around it in the field of measurement

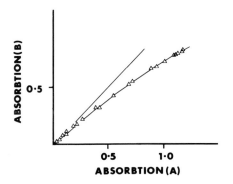

Figure 10.2 The relations between the absorptions of homogeneous and inhomogeneous nuclei at each of the wavelengths measured for the construction of the curves shown in Figure 10.1

stained material. Inhomogeneity of the field will cancel out in the final calculations.

In practice the sequence of operations for measuring Feulgen dye contents of nuclei with Patau's two-wavelength method is as follows. First, a specimen is prepared in which there are some homogeneously stained objects. Large cell nuclei such as those of newt or salamander erythrocytes pretreated with distilled water or a hypotonic salt solution for a few minutes and then fixed in 3 : 1 ethanol–acetic acid serve well for this purpose. The nuclei are stained with Feulgen and one of them is used to construct an absorption curve of the kind shown in Figure 10.1. This means arranging that the measuring area of the microdensitometer is entirely filled by a part of a stained nucleus. Two wavelengths are then selected such that the absorbance at the first wavelength is exactly half the absorbance at the second (absorbance $= 1/\log t$, where t = percentage transmission). The absorbances of the inhomogeneously stained specimens are then measured at both the selected wavelengths, and the Feulgen dye content of each specimen is calculated using Patau's formulae and tables (Patau, 1952; Pollister *et al.*, 1969). The method is slow and demands very careful and consistent microscopical technique, but as a rule the results are accurate and dependable.

The second method for coping with inhomogeneous specimens is much simpler for the operator but much more expensive in terms of equipment. It involves using a very small measuring area in the form of a spot of light (546 nm wavelength for Feulgen), setting the instrument to zero absorbance or 100% transmission by placing the spot over an empty area of the preparation, and then causing the spot to scan the specimen either by moving the specimen in relation to the spot or by moving the spot in relation to the specimen. Machines that are capable of measuring Feulgen dye contents in this way scan each specimen and then integrate the total absorbance encountered by the spot in the scanned area. Scanning and integrating microdensitometers that employ a moving stage and a stationary spot are manufactured by Carl Zeiss and by Leitz. Some of this equipment is exceedingly sophisticated and powerful: it is also quite expensive. A highly successful, thoroughly effective and somewhat less costly machine that employs a moving spot is manufactured by Vickers Instruments. Details of this range of equipment can be obtained from the manufacturers. Scanning and integrating microdensitometers can cope with high levels of inhomogeneity in staining and they are fast and simple to use. However, it has to be stressed again that even the best of them is only as good as its operator!

II. The Feulgen reaction

A. Fixation

It is important to keep to simple fixatives and if at all possible to avoid aldehydes. The Feulgen technique, after all, depends on the interaction of aldehyde groups with a colourless derivative of the dye basic fuchsin, so it makes sense to avoid or at least control the use of aldehydes as fixatives. The most faithful fixative for Feulgen is 3 : 1 ethanol-acetic acid. Fixatives that include formaldehyde are quite acceptable so long as the preparation is copiously washed after fixation to remove free formaldehyde and some control preparations that have not been hydrolysed (see Section II.B below) are processed along with the ones that are to be measured. The unhydrolysed controls should show no colour whatever.

The optimal fixation time in ethanol-acetic acid or any other fixative should be determined by preparing a series of slides of material that has been in fixative for different times and measuring the Feulgen dye contents of each. That which gives the highest value has been fixed for the optimal time.

As a rough guide, blood smears or small blocks of tissue will require from 15 min to 2 h in freshly prepared 3 : 1. Fixation time in formaldehyde or in an ethanol-acetic acid–formaldehyde mixture (Serra's fixative) should be longer, between 1 and 5 h.

After fixation in ethanol–acetic acid or in Serra's mixture the tissue should be transferred through several changes of 95% ethanol and then taken down through 70% and 50% ethanol to water. Tissue fixed in formaldehyde should be well washed in water or dilute buffer.

B. Hydrolysis

This is certainly the most important step in the Feulgen reaction. Hydrolysis is usually carried out with HCl. The strength of the acid and the time of hydrolysis are both quite critical. The relationship between hydrolysis time and staining intensity for three different sets of conditions are shown in Figure 10.3. From these graphs it will be evident that when employing 1 N HCl at 60°C the hydrolysis time is extremely critical: a few minutes either side of the optimum and the staining intensity drops away quite quickly. The 3.5 N and 37°C system and the 5 N and 20 °C system are much less critical and allow a broad time span over which staining intensity remains at maximum.

As with fixation, the time and conditions of hydrolysis *must* be determined experimentally at the start of each and every programme of microdensitometry.

Optimal hydrolysis time will be different for different tissues and fixatives.

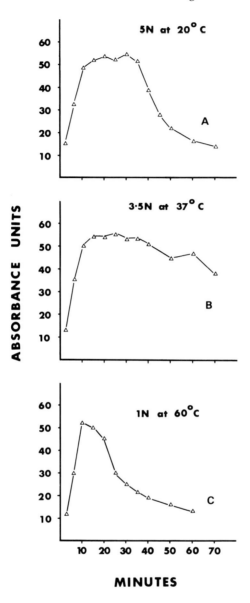

Figure 10.3   Curves to show the effect of hydrolysis time on the amount of Feulgen dye bound per cell nucleus. It should be clear that in curves A and B, where hydrolysis is in 5 N HCl at 20°C and in 3.5 N at 37°C respectively, hydrolysis time is not too critical between 10 min and 30 min; whereas in 1 N HCl at 60°C (curve C) hydrolysis time is quite critical and will produce sub-optimal staining if only slightly more or less than 10 min duration

After hydrolysis the tissue must be washed thoroughly in water to remove all traces of acid.

C. Staining

Schiff's reagent is prepared by the reaction of SO_2 with an acidic solution of basic fuchsin. It is entirely colourless. When Schiff's reacts with aldehyde groups it quickly becomes coloured. The Schiff's reagent–aldehyde complex is not the same colour as the original basic fuchsin solution. The two have quite different absorption spectra and are therefore considered to be different dyes.

Schiff's reagent is made up as follows:

(1) Add 0.5 g of basic fuchsin to 100 ml of water at room temperature.

(2) Add 1 g of potassium or sodium metabisulphite and 10 ml of 1 M HCl.

(3) Leave standing and shake from time to time until the solution has lost most of its colour and is clear and yellowish (straw-coloured) in appearance. This may take several hours.

(4) Add 0.5 g of activated charcoal powder, shake, and filter. The filtrate should be perfectly clear and colourless. Store in a tightly stoppered bottle in the refrigerator.

Yet again, the time of staining in Schiff's reagent is critical. The dye solution is quite acid and hydrolysis therefore continues slowly during staining. After a time DNA will actually be lost and staining intensity will decline accordingly. As a general rule, staining should not be longer than 2 h, but a maximum can be reached with some tissues after as little as 20 min. The answer, as with fixation and hydrolysis, is to determine the optimal staining time by experiment.

D. Washing

After staining the tissues or slides should be washed in several changes of SO_2 water over a period of 30 min. This removes all of the coloured Feulgen dye that is not bound to DNA. It is a very important step but not critical in terms of timing. The same washing procedure should suffice for all preparations: one quick rinse to remove most of the Schiff's reagent carried over from the staining dish, followed by three washes of 10 min each.

SO_2 water is made up as follows: 10 ml 1 N HCl; 10 ml 5% sodium or potassium metabisulphite; 180 ml water. It should be made up fresh each time it is needed. It should smell strongly of SO_2.

E. Dehydration and mounting

If tissue has been stained prior to squashing or splashing then transfer it to the appropriate solution – 45% acetic acid for squashes or 3 : 1 ethanol (or methanol)–acetic acid for splashes – and make preparations in the normal way (see Chapter 3, Section III, for squashes and Chapter 2, Section I, for splashes).

  Preparations should then be dehydrated carefully through 95% and 100% ethanol and cleared in several changes of good quality xylene. They should finally be mounted in immersion oil and covered with coverglasses of the thickness recommended for use with the objectives on the microdensitometer. Preparations can be mounted in balsam or in a synthetic mountant if necessary, but some of these may reduce staining after a time and some are less well matched to the tissue with regard to refractive index than is immersion oil. The principal advantage of using immersion oil is an optical one. Particularly if an oil-immersion objective is used on the microdensitometer then light will pass from the condenser through immersion oil to the slide, from the slide through immersion oil and the specimen to the coverglass, and from the coverglass through immersion oil to the objective. Under such conditions, and with the condenser properly positioned, the amount of light scatter will be reduced to an absolute minimum, and with some microdensitometric systems this is important.

F. Measuring

The preparations are now ready to measure. There is very little that can usefully be said about procedures for measurement since each different microdensitometer will require quite a different procedure and, according to the design of the instrument, different aspects of the measuring procedure will be more or less critical. However, there are certain general rules and comments that may prove helpful.

  First, be sure that the preparations are of good quality with regard to fixation and staining. Badly fixed nuclei will most surely bind Feulgen dye in amounts that are not at all related to their DNA content. Next, be exceedingly fussy about the condition and arrangement of the instrumentation: clean objectives, clean and properly adjusted eyepieces so that there will be no problem in obtaining a sharp focus and a clear image. Next, make many measurements at first to test reproducibility. Measure the same specimen time and time again, going away from it, altering the microscope, coming back again, resetting, and remeasuring. The result should be the same each time. Work at it until it is. If there is variation then tune the instrument and the measuring procedure until the readings come out with monotonous reproducibility! This will take time. For example, working with a Zeiss MPM and the two-wavelength method or with the Vickers M85 scanning and integrating machine it may take as long as 2–3 hours and over 100 measurements before the operator's technique becomes sufficiently uniform and the machine well enough tuned to produce consistent readings. It is often valuable to look for opportunities to test the instrument and the measuring procedure on specimens of known DNA content. Testis, for example, provides an excellent system for this purpose. Spermatids are easily recognizable and are absolutely known to contain the 1*C* amount of DNA. These can be compared in the same preparation with, say, pachytene nuclei that are equally well known to have the 4*C* amount of DNA, or with somatic nuclei that will have the 2*C* amount. Such differences should be clearly reflected in the microdensitometric readings if staining, instrumentation, and operator technique are good.

Finally, a word about the kinds of nuclei and other objects that can be measured and the interpretation of results. With regard to determination of *C* values there is no doubt that the erythrocyte nuclei of birds, reptiles, and amphibians are particularly useful. Just make a blood smear, fix it, and stain it. A sample protocol for the erythrocytes of certain plethodontid salamanders is given in Mizuno and Macgregor (1974). In other organisms, testis or liver are useful in the sense that they are generally easy to fix, stain, and squash. Smaller objects, such as individual chromosomes or parts of chromosomes, can be measured quite easily given a system of apertures that allows the field of measurement to be reduced to sensible proportions in relation to the size of the object to be measured. In practice, this means employing an aperture that is small enough to fit closely around the object whilst at the same time not at any point cutting into that object and preventing a part of it from contributing to the absorbance.

G.   Calibrating

The product of a microdensitometer is an absorbance value in arbitrary units. To convert these units to absolute values for amounts of DNA per nucleus or per chromosome it is necessary to compare the values obtained from an unknown material with those obtained under identical conditions from nuclei or chromosomes for which the DNA content has already been established. By far the best procedure is to include on the same microscope slide nuclei with known and unknown DNA contents, both fixed and prepared in exactly the same way. The two sets of nuclei are then unavoidably processed through the Feulgen reaction and prepared for microdensitometry in exactly the same way and the readings taken from them are likely to be strictly comparable. The DNA contents of certain objects are very well known indeed, and if at all possible one of these should be used as a reference standard in any programme that requires absolute DNA amounts. For example a human diploid cell in G_1 before it has begun replicating its chromosomes in preparation for the next mitosis has 7 pg of DNA. An erythrocyte nucleus from *Xenopus laevis* has 6 pg of DNA. A sperm nucleus from *Drosophila melanogaster* has 0.14 pg of DNA. These are particularly widely accepted values. There are, of course, many other well known *C* values: those for over 500 species of animal have been measured by microdensitometry, but a truly accurate estimate of nuclear DNA content can only be obtained by combining values obtained from a wide range of techniques and then deciding on a reasonable compromise. This has only really been done for a handful of animal species of which man, *Xenopus* and *Drosophila* are perhaps the best known and the most readily accessible.

The very latest advances in microdensitometry and microscope image analysis can be found in some of the highly sophisticated equipment currently produced by Carl Zeiss and Ernst Leitz. For instruments such as the Zeiss IBAS Interactive Image Analyzer or the Microvideomat 3 or the Leitz TAS Plus or MPV 3 – essentially high quality light microscopes surrounded by boxfuls of microprocessors, computers, and elaborate circuitry – the measurement of Feulgen dye contents of individual human chromosomes presents no problem, and the operators can amuse themselves by asking the machine to do such things as scanning along the length of a Giemsa banded chromosome, enhancing the image to clarify the banding pattern on a video screen, and then providing a graphic print-out of band densities and distributions. Technology like this, even if it costs the Earth, is truly remarkable and represents the present-day sum of all the contributions and accomplishments of generations of cytologists, microscopists, physicists, and engineers.

References

Deitch, A. D. (1966). Cytophotometry of nucleic acids. In *Introduction to Quantitative Cytochemistry* (G. L. Wied, ed.), Academic Press, New York and London, pp. 327–354.

Feulgen, R., and Rossenbeck, H. (1924). Microskopisch-chemischer Nachweis einer Nucleinsaure von Typus der Thymonucleinsaure und die darauf beruhende selective Farbung von Zellkerner in mikroskopischen Praparaten, *Z. Physiol. Chem.*, **135**, 203.

Kasten, F. H. (1960). The chemistry of Schiff's reagent. *Internat. Rev. Cytol.*, **10**, 1.

Kasten, F. H. (1970). The potential of quantitative cytochemistry in tumor and virus research. In *Introduction to Quantitative Cytochemistry*, II (G. L. Wied and G. F. Bahr, eds), Academic Press, New York and London, pp. 263–296.

Melamed, M. R., Mullaney, P. F., and Mendelsohn, M. L. (1979). *Flow Cytometry and Sorting*, John Wiley and Sons, New York.

Mizuno, S., and Macgregor, H. C. (1974). Chromosomes, DNA sequences, and evolution in salamanders of the genus *Plethodon. Chromosoma (Berl.)*, **48**, 239–296.

Patau, K. (1952). Absorbtion microphotometry of irregular shaped objects. *Chromosoma (Berl.)*, **5**, 341–362.

Pollister, A. W., Swift, H., and Rasch, E. (1969). Microphotometry in visible light. In *Physical Techniques in Biological Research*, III (A. W. Pollister, ed.), Academic Press, New York and London, pp. 201–251.

Swift, H. (1955). Cytochemical techniques for nucleic acids. In *The Nucleic Acids*, Vol. II, (E. Chargaff and J. N. Davidson, eds), Academic Press, New York, pp. 51–92.

Appendix 1

Names, addresses, and telephone numbers of suppliers

Animal suppliers

Bioserv Ltd, 38–42 Station Road, Worthing, Sussex BN11 1JP, England. Telephone: (0903) 200844.

Carolina Biological Supply Company, Burlington, North Carolina 27215, USA. Telephone: (919) 584 0381.

Centre for Reptile and Amphibian Propagation, 8166 East Shaw Avenue, Clovis, California 93612, USA. Telephone: (209) 298 6423.

Griffin and George, Gerrard Biological Centre, Gerrard House, Worthing Road, East Preston, West Sussex BN16 1AS, England. Telephone: (090 62) 72071.

Lee's Newt Farm, 107 Cooper Circle, Oakridge, Tennessee 37830, USA. Telephone: (615) 482 2638

Philip Harris Ltd, Biological Division, Oldmixon, Weston-super-Mare, Avon BS24 9BJ, England. Telephone: (0934) 413063.

Xenopus Ltd Biological Suppliers, Holmesdale Nursery, Mid Street, South Nutfield, Redhill, Surrey RH1 4JY, England. Telephone: (073 782) 2687.

Biochemicals and reagents

British Drug Houses (Chemicals) Ltd, Broom Road, Parkstone, Poole, Dorset BH12 4NN, England. Telephone: (0202) 745520.

British Drug Houses (US distributor): Gallard Schlesinger Chemical Mfg Corp., 548 Mineola Avenue, Carle Place, New York 11514, USA. Telephone: (516) 333 5600.

Bio-Rad Laboratories, 2200 Wright Avenue, Richmond, California 94804, USA. Telephone: (415) 234 4103.

Bio-Rad Laboratories Ltd, Caxton Way, Watford, Hertfordshire WD1 8RP, England. Telephone: (0923) 40322.

Boehringer Mannheim, BCL., The Boehringer Corporation (London), Bell Lane, Lewes, East Sussex BN7 1LG, England. Telephone: (079 16) 71611.

Boehringer Mannheim Chemicals, 7941 Castleway Drive, PO Box 50816, Indianapolis, Indiana 46250, USA. Telephone: (317) 849 9350.

Difco Laboratories, PO Box 14B, Central Avenue, West Molesey, Surrey KT8 0SE, England. Telephone: (01) 979 9951.

Difco Laboratories, PO Box 1058A Detroit, Michigan 48232, USA. Telephone: (313) 961 0800.

Enzo Biochemicals Inc., 325 Hudson Street, New York, N.Y. 10013, USA. Telephone: (800) 221 7705 (for callers in the USA); (212) 741 3838 (for callers from outside the USA)

Fisons Scientific Apparatus Ltd, Bishop Meadow Road, Loughborough, Leicestershire LE11 0RG, England. Telephone: (0509) 31166.

Gibco Europe Ltd, PO Box 35, 3 Washington Road, Paisley PA3 4EP, Scotland. Telephone: (041) 887 6111.

Gibco Laboaratories, PO Box 68, Grand Island, New York 14072, USA. Telephone: (716) 773 0700.

Hopkin and Williams Ltd, Freshwater Road, Dagenham, Essex RM8 1QJ, England. Telephone: (01) 599 5141.

Miles Laboratories Ltd, PO Box 37, Stoke Court, Stoke Poges, Buckinghamshire SL2 4LY, England. Telephone: (028 14) 5151.

Miles Laboratories, PO Box 2000, Elkhart, Indiana 46515, USA. Telephone: (219) 264 8804.

Oxoid Ltd, Wade Road, Basingstoke, Hampshire RG24 0PW, England. Telephone: (0265) 61144.

Oxoid Canada Ltd, 145 Bentley Avenue, Ottowa, Ontario K2E 6T7, Canada. Telephone: (613) 836 5386.

Sigma London Chemical Company, Fancy Road, Poole, Dorset BH17 7NH, England. Telephone: (0202) 733210.

Sigma Chemical Company, PO Box 14508, St Louis, Missouri 63178, USA. Telephone: (314) 771 5750.

Union Carbide (UK) Ltd, Union Carbide House, Chemical Division, 95 High Street, Rickmansworth, Hertfordshire, England. Telephone: (0923) 720166.

Materials and Equipment

American Scientific Products, 1430 Waukegan Road, McGaw Park, Illinois 60085, USA. Telephone: (312) 689 8410.

Dow Corning Corporation, Midland, Michigan 48640, USA. Telephone: (517) 496 4000.

ASCO Laboratories, 52 Levenshulme Road, Gorton, Manchester M18 7NN, England. Telephone: (061) 223 9343.

A. R. Horwell Ltd, 2 Grangeway, Kilburn High Road, London NW6 2BP, England. Telephone: (01) 328 1551.

R. A. Lamb Ltd, 6 Sunbeam Road, London NW10 6JL, England. Telephone: (01) 965 1834.

Slaters Plasticard Ltd, Royal Bank Buildings, Temple Road, Matlock Bath, Matlock, Derbyshire DE4 3PG, England. Telephone: (0629) 3993.

Arthur H. Thomas Company, PO Box 779, Philadelphia, Pennsylvania 19105, USA. Telephone: (215) 574 4500.

Whatman Laboratory Sales, Ltd, Springfield Mill, Maidstone, Kent ME14 2LE, England. Telephone: (0622) 674821.

Radiochemicals

New England Nuclear, 549 Albany Street, Boston, Massachussets 02188, USA. Telephone: (617) 482 9595.

Amersham International PLC, White Lion Road, Amersham, Buckinghamshire HP7 9LL, England. Telephone: (024 04) 4444.

Photographic and Autoradiographic

Ilford Sales Company, 14–22 Tottenham Street, London W1, England. Telephone: (01) 636 7890 (orders) and (0565) 50000 (information).

Eastman Kodak Company, Rochester, New York 14650, USA. Telephone: (716) 325 2000 (orders) and (716) 724 4633 (information)

Kodak Limited, Kodak House, Station Road, Hemel Hempstead, Hertfordshire HP1 1JU, England. Telephone: (0422) 61122.

Computing and Data Analysis

The Leicester Computer Centre Ltd, 67 Regent Road, Leicester LE1 6YF, England. Telephone: (0533) 556268.

Index

DISCARD